A CENTURY OF
ROAD MATERIALS

Frontispiece (overleaf) *Night-tipping of molten slag at Teesport*

A CENTURY OF ROAD MATERIALS

*The history of
the Roadstone Division
of Tarmac Ltd*

J. B. F. EARLE
B.Sc., D.I.C.

BASIL BLACKWELL · OXFORD

0 631 13890 0

Printed in Great Britain by
The Camelot Press Ltd, London and Southampton
and bound by Kemp Hall Bindery, Oxford

Preface

When Tarmac, Derbyshire Stone and William Briggs merged in the middle of 1968, the enlarged Tarmac renamed Tarmac Derby Limited,* which was chosen as the vehicle of the merger, became not only the largest slag entrepreneur in the United Kingdom, but much the largest roadstone quarrying enterprise and the largest road asphalt manufacturer.

Since 'black-top' constitutes 80 per cent of the road surfaces in the United Kingdom, and since rock, gravel, or slag form nearly all the foundations, the history of the Roadstone Division of Tarmac Derby affords a view of a significant and typical slice of the major part of the road materials industry.

This is a considerable industry, and there is a dearth of economic study of it. I hope, therefore, that the contents of this book may be of interest to a wider circle than that of the executives of Tarmac Derby Ltd., for whose edification it was originally projected.

Original documents relating to the Division's constituents are scanty. Directors' minute-books have largely been preserved but their contents are frequently highly laconic, and they rarely give much guidance as to motive. Propaganda booklets have been a valuable source, but from their very purpose, have had to be regarded with caution. Annual accounts are sometimes useful, but are frequently so confused with diversifications as to be very

* While the book was in the press, the company's name was changed back to Tarmac Ltd.

misleading. This is particularly the case where civil engineering activities, which my brief excludes, are involved.

I have carried the story up to the merger in 1968, though a certain reticence has been necessary in the last few years. Many of the constituents seem to have first taken shape in the second half of the nineteenth century, which provides my excuse for the title which I have chosen.

My task would have been impossible without the facility with which I have been able to draw on the memories of senior and in particular retired members of the staffs of the various companies. My gratitude for the co-operation and heartfelt interest of these gentlemen is immense. I have given them a great deal of trouble, and they have rewarded me with hospitality and kindness.

My warm thanks are likewise due to the officials of the various public bodies whom I have had occasion to consult, and especially for the run of their specialized libraries. I would mention in particular the British Road Federation, the Iron and Steel Institute, the Institute of Quarrying, the Institute of Asphalt Technology, the Asphalt and Coated Macadam Association, and the British Quarrying and Slag Federation.

The University of Lancaster was good enough to appoint me an honorary Research Fellow largely in order to facilitate my writing of this book, and I would like to express my appreciation to the Vice-Chancellor and to Professor P. W. S. Andrews and the staff of the Department of Economics for their most friendly help in every direction.

I have received much help from friends connected with the industry, though not directly with the Tarmac Derby Group, and would especially mention Mr. Rowland Green of Stothert & Pitt Ltd., whose guidance on plant development has been invaluable.

I should mention four points of stylistic detail. I have frequently used the shortened names by which the constituent undertakings, and other well-known companies, are or were commonly known. I include in the expressions 'roadstone', 'road materials', etc., similar materials used for other forms of construction, such as airfields. I use the spelling 'asphalte' only for the material derived from natural sources; otherwise I spell the material 'asphalt'. I use the word 'granite' in its popular sense, not in its precise petrographical meaning.

There are a bewildering number of places mentioned in the book. I have tried to help readers by providing a special index with county locations.

J. B. F. EARLE
London, 1970

Contents

Illustrations

Introduction

Roadstone is a word lacking in romantic or emotional overtones. Yet the development of our country in this century of extraordinary scientific and technological advance has depended to such a degree on the utilization of the internal combustion engine in road vehicles that those engaged in the roadstone industry are entitled to regard themselves as contributing to the essential infra-structure of our social and commercial existence. The same, of course, applies to the motor industry, but it lacks that sense of deriving directly from our natural resources, which roadstone shares with coal and agriculture. This book is about the largest unit in a fundamental activity of life in Britain.

In reading it, it should be understood that when the Roadstone Division was constituted after the 1968 merger, its activities were intended to be confined to the United Kingdom, the overseas activities of the Tarmac Derby Group being normally controlled by different organizations. For all that, the story carries us abroad from time to time. The road asphalt business in this country was founded on the importation of rock asphalte from the mines of France and Switzerland, while the ramifications of the trade in sea-borne roadstone are so wide and to my mind, so interesting, as to merit the inclusion of a special section in Chapter VIII.

My story is one with many threads—there are twenty separate narratives in Chapters II, III, and IV—but it has three main strands. The enterprises which constitute the Division divide themselves into: (i) those primarily concerned with the exploitation of blast-furnace and other slags; (ii) those primarily concerned in the production and marketing of natural stone, and: (iii) those concerned in the manufacture and laying of road asphalt.

For fifty years exploitation of slag in Great Britain consisted largely in coating the aggregate with tar or bitumen in order to produce road materials demanded by the automobile age. The major use of the quarries in the Group was likewise as road material. Road asphalt consists of bitumen combined in some way with aggregates derived from stone or slag. It would be pedantry to exclude from the expressions 'roadstone', 'road materials', and 'road asphalt' practically identical materials used for such purposes as the construction of airfields, footpaths, playgrounds, car parks and tennis-courts. Consequently the Division has the most appropriate short title that can be devised. It must, however, be made clear from the start that the Division and its constituents are and have been in no way confined to the production and marketing of road material.

Chapter X therefore describes non-roadmaking activities. Every aggregate, slag, igneous and sedimentary rock, limestone, and gravel has found substantial uses outside roads. Limestone, though, is a chemical raw material, and as such enters into many another field. There are extraneous activities connected with slag, one of the most important being metal reclamation at steel furnaces. The mining of non-ferrous metallic ores has been not insignificant.

Nevertheless, this is primarily the history of the constituents of the country's largest producer of road materials, of their ramifications, their combinations, their successes and their failures. It cannot be understood without a comprehension of the background of demand for roads which has been growing, save for the periods of the two World Wars, ever since the twentieth century began. Even before plunging into the narratives of the companies, therefore, I have written Chapter I—'The demand for road materials'.

The history ends in 1968, on the conclusion of the merger. It seems that a convenient starting date is about 1870. Stone has of course been quarried for road purposes from time immemorial, and at least three of the Division's quarries, Caldon Low, Hopton Wood, and Bankfield at Clitheroe, have histories dating back to the eighteenth century. The narratives will, however, show the majority of the quarries emerging from the background of a land-owner's or farmer's sideshow in the last thirty years of the nine-teenth century.

Similarly, although the rock asphalte in the canton of Neuchatel was discovered as early as 1712, road asphalt was first laid on a London carriageway in 1869. Slag comes a little later. Purnell Hooley's discoveries were made in 1901.

After the narrative chapters, I have in Chapters V, VI, and VII, sought to weave the threads into the three strands of slag, of quarrying and coated macadam generally, and of road asphalt and surfacing. It will be found that the slag and quarrying themes differ radically from one another, though the end products are sold in the same markets. Chapter VII, with the fundamental change in binder from European rock asphalte to the residual bitumen of the oil companies, and the startling effect of the introduction of the mechanical spreader, will, I hope, be found particularly interesting. These three chapters inevitably include a certain amount of technical matter, since one must trace the profound effects of development of plant and of end-product specification.

Chapter VIII is about transport. It is at this point that I must say something about the geological and geographical background against which the whole book is written. Where the fundamental raw materials are to be found, whether they be igneous and sedimentary rock, limestone, gravel and sand, iron ore and coal (together with limestone the necessary ingredients of iron-making) or natural asphalte has determined the whole course of the century's development.

'South-east of the line Lyme Regis to King's Lynn, in which a third of the population of Great Britain lives, there is no igneous rock being quarried and very little limestone, except for the Kentish Ragstone of the Maidstone district, which, since it alternates with a clayey sandstone, known locally as hassock, is expensive to work.' These words, quoted from my monograph on English

Foreign Trade in Roadstone, convey an important influence on the present history. It may be added, neither is there any slag production in that area except for about 100,000 tons per annum at the Ford Motor Co. Ltd. at Dagenham. What there is, however, in this region of tertiary geology, is plenty of gravel, and the Roadstone Division does not possess any, except at Cray River in Kent. Two of its constituents, Chittenden & Simmons and Crow Catchpole, once had Kentish ragstone, but disposed of it long ago. However, in a market requiring perhaps 8,000,000 tons of aggregate per annum, there is a great demand for slag, igneous rock, and hard limestone, and frequent reference will be found in this history to the Division's constituents' efforts and successes in supplying a large proportion of the needs of this area.

The igneous and sedimentary rocks and the hard limestones all lie to the north and west of the island, some near the sea coast. The latter will figure in Chapter VIII. Scotland has little hard limestone, but possesses great reserves of whinstone, widely distributed. It also has plenty of good gravel, especially in the Forth/Clyde industrial belt. Moving south, we meet the whinstone of northern England, with the narrow bands of the Whin Sill and the Whin Dyke, both originating at Cross Fell, the one reaching the sea at Bamburgh, the other at Whitby. The major source of roadstone, however, not only in the north of England, but taking England as a whole, is hard limestone. The Pennines and their foothills and the connected Lake and Peak Districts are full of good limestone, with occasional deposits of igneous or sedimentary rock, as at Waterswallows, Taddington, and Arcow. But the south-easterly parts of the north Midlands—using this expression in its widest extent—have very few roadstone quarries, and Lancashire is singularly bare south of the Ribble. Gravel occurs nearly everywhere, but there is little other stone suitable for first-class road materials in the East Riding, the area of conurbation in the West Riding, in Nottinghamshire, Lincolnshire or south Staffordshire. As one continues to move south, however, one strikes great deposits of igneous and sedimentary rock in Leicestershire, Warwickshire, Worcestershire and Shropshire. We then come to my King's Lynn–Lyme Regis line and to tertiary or even later geological systems.

E. PURNELL HOOLEY

SIR ALFRED HICKMAN, BART., M.P.

There remain to be recorded in England the immense limestone deposits of the Mendips and south Gloucestershire, and the granites and limestones of the south-western peninsula. In Wales, there is plenty of igneous and sedimentary rock and some limestone in the north and centre, and South Wales is covered with limestone quarries.

The overwhelming importance of the transport factor flows from these geological circumstances. It is particularly noteworthy that the three greatest conurbations in England, Merseyside, the West Riding, and London, have in effect no indigenous roadstone except gravel. Small wonder, therefore, that the London market, for instance, has been supplied with aggregates as geographically diverse as German, Belgian, and French slag, Norwegian, Belgian, and Irish granites, Kentish ragstone, West Country limestone, British slag from almost every ironmaking district and Scotch, Welsh, Channel Islands, Cornish, Leicester, and Northumbrian igneous rock.

This is a business history, and the commercial policies of the Division's constituents are naturally very relevant to their fortunes. My remaining general subject therefore is set forth in Chapter IX—'Co-operation in the Industry'. I try to trace there the emergence from Victorian individualism and hostility to competitors. It covers a wide range of objects and methods—protective action against transport monopolies and Government interference, joint activity in research and propaganda, price associations and price reporting centres, joint ventures, joint companies, elaborating into mergers on a big scale, being among the many matters discussed. The book concludes with some general observations.

It has been said that 'History is about people'. The main fascination of compiling a study of this nature is in trying to assess the characters and motivations of the men who have been in a position to affect the fortunes of the businesses being described. The century which my story covers has seen profound changes in men's approach to the solution of industrial problems. Nowhere is this more obvious than in the quarrying industry. Indeed industry is a microcosm of life, and the social history of our times is illumined as we see the stages whereby the management of the exploitation of a hundred million tons of granite, as at Cliffe Hill, goes through all the stages from the dictatorship of a Victorian squire to professional administration versed in the disciplines of the modern Business School, and advised by experts in planning, accountancy, engineering, geophysics, and chemistry. The interesting and outstanding characters are not only on the Division's side of the fence. Customers have changed in nature and education just as much as managers. We must try to picture the Edwardian County Surveyor, frequently dominating his fellow country gentlemen who form his Highways Committee and compare him with the highly trained professional civil servants who are today the directors of Road Construction Units.

CHAPTER I

The Demand for Road Materials

Had you seen these roads before they were made
You would lift up your hands and bless General Wade.
Inscription on the Wade monument between
Inverness and Fort William

It is sometimes difficult to realize that the problem of inadequacy of roads, both in quantity and quality, is no brand-new twentieth-century problem. It would probably have been a considerably easier matter to provide a British road system suited to the automobile age if there had not existed a mass of road legislation and administration painfully derived from the age of horse-traction. Experience in undeveloped countries certainly seems to confirm this, as does the extraordinary speed of construction which characterized the building of the British railway system in a time of limited civil engineering technology. We must, therefore, have a brief look at the past. I have drawn so largely, for this purpose, on the works of Professor O'Flaherty[1] and of the late Mr. Rees Jeffreys[2] that I will omit specific references to their pages.

Roads started with the invention of the wheel in Mesopotamia over 5,000 years ago, and examples of substantial construction have been identified in Egypt, Crete, and Babylon. In the latter case bitumen was used for jointing purposes. The Persian Royal Road, built around 500 B.C. and the road constructed in India by

[1] C. A. O'Flaherty, *Highways*, Arnold, 1967.
[2] Rees Jeffreys, *The King's Highway*, Batchworth, 1949.

Chandragupta in 322–298 B.C. were each some 1,500 miles long. We know, however, of no formal roads in barbarous Britain prior to the Roman conquest of A.D. 45.

Roads represented high imperial policy to the Romans. The 372 great roads of the Empire extended to about 48,500 miles, and Britain (south of the Forth–Clyde line) had its fair share. The Romans' road network formed the core of the British road system from their departure about A.D. 410 until this century.

For the next fifteen hundred years, until the Road Board was established in 1910, there was no central road authority in Britain. Indeed for more than 500 years, there was no central authority at all. The coronation of King Edgar in 973 may be regarded as marking the establishment of the unified English state. Small wonder, therefore, that roads became the responsibility of local authorities, in effect of the parishes and boroughs, whose interest in through roads was very small. As the medieval era came to a close, British roads became worse and worse. The only systematic approach to road maintenance had come from the monasteries, who regarded it as a charity for travellers, and with their dissolution, it is probable that the state of the roads in the sixteenth and seventeenth centuries reached its nadir.

It was not till 1555 that the Act 'For amending the Highways being now very noisom and tedious to travel and dangerous to all Passengers and Carriages' was passed, and all it did was to place a statutory obligation on the parishes to maintain their roads by means of obligatory unpaid labour under the direction of an unpaid Surveyor of no qualifications. The ineffectiveness of the legislation led inevitably to the attitude that traffic was a nuisance that had to be suppressed if possible. An Act of 1621 forbade the use of four-wheeled wagons and the carriage of goods of more than one ton weight.

I may pause here to note the very early origin of the two antagonisms which have bedevilled roadwork, not least in the hundred years of our survey. One is the conflict over the roads between the local authorities and the central power; the other is the feud between the road-user and the road-maker.[1] Without a

_____
[1] J. W. Gregory, The Story of the Road, Alexander Maclehose, 1931.

historical background, it might well be believed that public roads must, through some ancient right in common law, be made capable of carrying whatever traffic is necessary for the welfare and progress of the locality. On the contrary, the victory of the ship-builders Barclay Curle & Co. in the House of Lords over Glasgow Corporation in 1923 must be regarded as modern law-making. Barclays had to transfer boilers weighing 80 tons along a public road between their works and their shipyard, and the Corporation sought to prohibit the traffic. The House of Lords held, however, that this was ordinary traffic of the district, and that the Corporation, as the highway authority, were required to provide a road that could carry it.[1]

As the eighteenth century opened, road-users accepted a new principle, that they, and not the general body of citizens, should pay for the upkeep of the roads. So Parliament enacted a succession of statutes creating the Turnpike Trusts. As many as 1,100 of these bodies, covering 23,000 miles of roads, were eventually consti-tuted, with powers of toll-raising and obligations of construction and maintenance. The demand for roads revealed the shortage of, and, at the same time created the supply of, skilled road engineers. Tresaguet, Inspector-General of Roads in France under Louis XVI, has been called the father of modern highway engineering, but in this country the great names are Robert Phillips, John Metcalf, the blind contractor of Knaresborough, Thomas Tel-ford and John Loudoun Macadam.

Telford (1757–1834) was a great civil engineer, but Macadam (1756–1836) was the first true highway engineering specialist. His engineering concepts are as valid today as they were 150 years ago. His two fundamental principles were:

(i) it is the native soil which really supports the weight of the traffic; while it is preserved in a dry state it will carry any weight without sinking;

(ii) To put broken stone on a road, which shall unite by its own angles so as to form a solid hard surface.

The Tarmac Derby Roadstone Division must regard itself as one

[1] Gregory, op. cit.

of Macadam's heirs, and applaud the imagination of the company's first chairman, Mr. E. P. Hooley, in devising the word 'Tarmac' to describe his new material.

Macadam was far more than a mere engineer. He really destroyed the prevalent hostility to increased traffic. A great administrator, he urged in particular that the surveyor must be a trained professional officer, of good social status, and paid a salary which enabled him to be above corruption.

During this turnpike era, the British Government's main direct intervention in roads was in Scotland. First, beginning in 1726, General Wade and his successors constructed some 800 miles of new roads in the southern Highlands thereby constituting a reversion to the Roman policy of dealing with the development and control of a backward and potentially rebellious province. Then in 1803–21, Telford constructed 920 miles of new roads for the Government in Scotland, which involved building 1,117 bridges.

Suddenly in 1835–50, the turnpike era ended. What Macadam described as 'the calamity of railways' fell upon the Turnpike Trusts, and bankrupted or terminated them. The rural roads thereupon reverted to the old system of parish maintenance and in 1850 there were some 15,000 separate and mutually independent highway authorities in England and Wales, in charge of roads practically deserted and considered in all probability to be useless.

The intolerable burden on the parishes compelled administrative reform. Under the Public Health Act of 1872, the Local Government Board assumed overall responsibility, and legislation culminated in the Local Government Acts of 1888 and 1894. The responsibility for rural main roads passed to the County Councils; the newly-created county boroughs were to look after all their own roads, while rural, urban, and borough councils took over the balance. Half the cost of main roads was borne by the Government, while block grants were made to the County Boroughs.

The big mistake of the Local Government Act of 1888 so far as highway planning and development were concerned was the separation of the County Boroughs from the administrative county. The cities had a great tradition of independence, and their streets had often advanced technically more rapidly than the rural roads. In the city of London, the Worshipful Company of Paviors has a history dating back to the thirteenth century, though it is noteworthy that it fell into desuetude throughout the first three quarters of the nineteenth. Asphalt paving was being rapidly extended in some of the cities by 1888; it was first introduced on a carriageway in London in 1869. But 1888 brought no co-ordination between the cities and the counties.

The second administrative error was the perpetuation of the rural district councils as highway authorities. As a result, there were still, after the reforms, some 1,800 highway authorities in England and Wales.

Nevertheless, it was very fortunate that a substantially improved financial and administrative structure had been introduced just in time for the motor age, which may be dated from the Locomotives on Highways Act of 1896.

With this historical background I can now turn to the subject of this chapter.

While cities all over the world, as the narrative of the Neuchatel Asphalte Company will show, were improving their main streets by paving them with rock asphalte during the last quarter of the nineteenth century, the pressure for improvement of rural roads came first from the cyclists. The advent of the 'Safety Bicycle' and of pneumatic tyres in the eighties caused a minor traffic explosion, and the new hobby spread like wildfire through all classes of society. In 1899 the Cyclists' Touring Club had a membership of 60,499. It was the cyclists who formed the Roads Improvement Association in 1886, and the newly-formed motoring organizations joined it. In 1901 the Association, by deputation and memorandum, tried to stir the Local Government Board into action, stressing in particular the need for a central road authority and for the payment out of national taxation of one half of main road maintenance in England and Wales. A departmental committee was set up and made some reformist recommendations, but none of them were implemented. An unsuccessful private members' Bill followed, of which Sir Alfred Hickman Bart. was one of the sponsors.

The failure of the Local Government Board is scathingly described by the famous Socialist economists, Sidney and Beatrice Webb, in *The Story of the King's Highway* published in 1913. However, developments forced the Government's hand, and the principal one was dust nuisance. Traffic dust was nothing new— in the coaching era one of Sheridan's characters tells of 'a tail of dust as long as the Mall' on the Bath Road—but with increasing motor traffic it became intolerable. Road engineers combined to defeat it by research and experiment and succeeded by tar spraying and by the laying of tarmacadam in place of water-bound material. Public pressure and well-directed propaganda in regard to roads now arrested the attention of the Chancellor of the Exchequer, Mr. Lloyd George, and provoked the remarks in his historic Budget speech of 1909 when he said (*inter alia*):

> It is quite clear that our present system of roads and of road-making is inadequate for the demands. . . . The State has for a very long period done nothing at all for our roads. . . . We propose to make a real start. . . . No main road has been made out of London for eighty years. We have no central road authority. . . . The Great North Road . . . is under 72 authorities . . . such as the Kirklington U.D.C. which controls one mile.

He proposed a new central authority, the Road Board, for the purposes *inter alia*, of making grants to local authorities to enable them to carry out well planned schemes for widening roads; for straightening them; and for making deviations round villages; for allaying the dust nuisance, and for constructing absolutely new roads. To finance his proposals, he set up 'The Road Improvement Fund' fed by proceeds of increased taxes on motor vehicles and a petrol tax imposed for the first time.

The Road Board was constituted in May 1910, the month in which King Edward VII died. As the new era opens, I must pause to try to relate the demands for road material in the previous forty years to the fortunes of the Tarmac Derby companies which were in existence (for the details of which the reader is referred to the appropriate sections of the next three chapters) and to try to visualize the atmosphere of the times in the road world.

Of all the companies, the only one of any financial status was The Neuchatel Asphalte Co. Ltd.; though its connection with British roads was only through its sales of natural asphalte to the Val de Travers Paving Co. throughout this period.

By 1910, road tar specialists like Catchpoles, and tar spraying companies like H. V. Smith were beginning to profit from the 'dustless roads' campaign.

As to the quarrying companies, Ord & Maddison were advertising whinstone and other road materials as early as 1869, and were early entrants into sett-making, as were Cliffe Hill. The battle between setts and asphalt as a paving for city streets was a very long drawn-out one, but it would appear from the narratives of companies such as Rowley Regis and Cliffe Hill that sett-making was not a particularly profitable line. Demand for broken stone and for setts did of course increase in the first decade of this century, with the local authorities' revenues being augmented by motor taxation, but it is difficult to deduce much advantage to profit. The Cliffe Hill figures slowly improved, probably because of their access by rail to the London market.

The most enterprising of these little concerns, seems to have been (New) Northern Quarries, and the illustration, facing page 63, shows their tarmacadam activity in Blackpool as early as 1904. James Ward appears to have shared with Purnell Hooley of Tarmac and William Prestwich a vision of the future which the coming of the motorcar presaged.

Tarmac was, of course, an imaginative conception. From its start in 1903, neither Hooley nor the Hickman family seem to have contemplated anything but black-top roads. Nevertheless, until 1911, very little profit was made. From every point of view, the establishment of the Road Board in 1910 marks a great turning point in this history, not least from its author's angle, since it is only from 1910 that we have any authentic statistics to guide us.

What manner of men were our predecessors' customers? The big money was spent by the County Councils. As will be seen from the story of the Road Board, Whitehall interference was practically non-existent. The County Councils' executive officers were the County Surveyors, and the Edwardian County Surveyors were interesting personalities. It must be remembered, too, that their professional lives lasted a long time, as they were generally appointed at an early age. Lt.-Col. J. P. Hawkins was County Surveyor of Berkshire from 1905 to 1947. The best of them complied very well with Macadam's specification, even as to the good social status. The County Councils had a high proportion of landed gentry in their membership, and the County Surveyor frequently appeared to the outside world as a country gentleman with a reasonable competence. His assistants frequently saw themselves in the same role with appropriate modifications.

I have ventured into this psychological field because it seems important to detect why some of the road material companies developed more rapidly and successfully than the others. One important reason seems to have been in the difference of their methods of salesmanship. In the narrative of Tarmac, I have described the first Managing Director, Mr. D. G. Comyn, as a great salesman. He himself, and he trained his sales staff to do likewise, succeeded in penetrating the social life of his customers. Of course, this had to be coupled with good material and good service, but many of Tarmac's competitors provided the latter without succeeding in making personal friends of their customers. It is noteworthy that when, at a later date, the road contractors became large customers, the Tarmac sales organization found it difficult to adjust itself to a purely commercial outlook, as it also did to Whitehall influence through the Divisional Road Engineers.

An interesting sidelight on the mixed motives of the customers of the Edwardian road materials industry is contained in a verbatim report in a local newspaper of the proceedings of the Roads and Bridges Committee of the local County Council in December 1902 which reads:

It being proposed to obtain Quenast (Belgian) granite, Mr. G. said it was a beastly shame, for the sake of 2d. to go in for foreign granite prepared by foreign labour.

Mr. B. suggested that the qualities of the granite should be tested by sections being put down lengthways with the different granites.

Mr. W.: I think the best test is to enquire which Company gives the best dinners. (Laughter)

Mr. D.: There is something in that.

Mr. M.: We thought Forest Rock gave a good dinner last year.

Mr. P.: This foreign company does not give dinners.

The conclusion was to split the order. The difference was in fact 1s. 8d. per ton.

In these early days, there were two other main classes of customer—the County Boroughs and the Rural District Councils. In the boroughs a lot of attention had to be paid to local councillors, who interfered in day-to-day management in a way rarely seen in rural areas. So far as the Rural District Councils were concerned, the situation became increasingly unsatisfactory. They clashed with the County Councils, and their roads were administered by badly underpaid and often far from competent surveyors. The Local Government Act 1929, which made County Councils in England and Wales the highway authorities for all roads in rural areas and classified roads in boroughs and urban districts was a very welcome measure of reform.

The Road Board of 1910 proved a disappointment to road-users, and when it was dissolved in 1919 no one mourned its passing. It was unpopular with the House of Commons, which disliked its financial independence, and the fact that only a Treasury minister could be questioned about its operations, for which the Treasury had no administrative responsibility. The Treasury appointed its members, and chose an ex-railway manager, Sir George Gibb, as Chairman; the Board in fact acted as a 'one-man-show', and Sir George Gibb confined its activities very much to the improvement of road crusts, though a few major road schemes were supported, mainly around London. The total money spent on roads did increase from £14·3m. in 1910 to £18·1m. in 1914[1] during which period it is to be noted that Tarmac Ltd.'s profit doubled. Since the Board originally decided not to appoint any permanent engineering staff, it had little influence on the highway authorities.

[1] *Basic Road Statistics, 1969*, British Road Federation, London.

In 1913 however Mr. (afterwards Sir Henry) Maybury, then County Surveyor of Kent, was appointed Chief Engineer; within a year World War I had started, road expenditure was cut to ribbons (only £12·2 m. was spent in 1917) and before long Maybury was in France in charge of roads for the Army. Though profits seriously declined, Tarmac weathered the storm by obtaining large military contracts, but a company like (New) Northern Quarries was forced into liquidation.

In 1919 the newly formed Ministry of Transport took over the powers and duties of the Road Board. Roads were placed under a separate department, headed by Sir Henry Maybury as Director-General.

Maybury was outstanding as both administrator (he set up seven divisions each under a Divisional Road Engineer), and engineer, and road work began to be tackled seriously. Vehicle duties were, in effect, all credited to the Road Fund—the motor spirit duties being abolished in 1920—and to relieve unemployment £1¼m. was specially contributed by the Treasury, who also made available £5¼m. for loans to local authorities. The result was that total expenditure on roads rose from £13·6m. in 1919 to £59·1m. in 1926. In real terms this meant that more than twice the work being done in 1914 was being done in 1926. The motor vehicle population had increased fourfold, but of course horse-traction had substantially diminished.

The period 1920–5 was a good one for the Tarmac Derby companies, as the narratives will indicate, but trouble was at hand. It started with Mr. (afterwards Sir Winston) Churchill raiding the Road Fund to solve his problems as Chancellor of the Exchequer, and succeeding Budgets continued the process until by 1936 the Road Fund was a dead letter, and there was no connection between motor taxation and the sums spent on roads.

Sir Winston Churchill frankly avowed that he did not want to see motor transport grow too quickly, and wanted to protect the railways, an excellent example of the feud between road-maker and road-user to which I have already referred.

A Royal Commission on Transport was established in 1928, and reported in 1931, when it advocated the building of by-passes and the reconstruction and improvement of existing roads, while rejecting the expenditure of large sums of public money on new arterial roads. Most, however, of its limited recommendations were ignored by the Government, who were by then plunged into financial crisis.

'The great depression' is generally dated from 1929, and it certainly lasted till 1935. The effect on total road expenditure should be studied in detail alongside the motor vehicle population. The figures[1] are:

Year	Expenditure £m.	Vehicles m.
1926	59·1	1·7
1927	58·6	1·9
1928	58·7	2·0
1929	57·1	2·2
1930	65·4	2·3
1931	66·3	2·2
1932	68·8	2·2
1933	54·0	2·3
1934	50·8	2·4
1935	51·5	2·6
1936	54·7	2·8

We see two things from these figures. The first is that even without the goodwill of the Government, and in the face of intense railway propaganda—'the Fair Deal', and in the midst of industrial depression, the vehicle population increased by more than 50 per cent in ten years.

The second is the delay between decisions of financial policy and their effect on actual road expenditure. The same effect can be seen in the opposite direction in 1919–23. The policy was settled in the former year when only £13·6m. was spent on the roads. It took until 1922 for the figure to reach £49·3m. The narratives which follow will indicate the critical situation of nearly all the companies in the period 1926–35, and how things became desperate in 1933 with a cut in expenditure of over 20 per cent.

[1] *Basic Road Statistics*, op. cit.

This prolonged period of shortage of demand had a profound effect on our companies. It had much to do with the change in iron company policy from direct marketing of slag to the employment of specialist entrepreneurs. It forced a complete change of attitude from Victorian individualism to industrial co-operation. The narratives will give one example after another of this, as men like Cecil Martin and John Hadfield (and Sir Henry Maybury, who left the Ministry of Transport in 1929 to become Chairman of British Quarrying Co. Ltd.) led the way to a new conception of the road materials industry.

The depression slowly lifted, under the influences of a great housing drive,[1] itself helpful to demand for the Tarmac Derby companies' products, and of rearmament as the German menace became apparent, but it was not until 1938 that the total expenditure on roads returned to the figure of 1930. Profits gradually improved. Then came World War II.

No statistics are available for road expenditure in the war years, but it fell to a very low ebb, and in 1945 was only £36·9m.

For many of the companies, however, and especially for the specialists in 'black-top', and those with civil engineering connections, it was a period of great demand for military purposes, airfield construction predominating. The catastrophic defeat on the Continent, followed by the Battle of Britain and the development of an aggressive strategy of heavy bombing of the countries under Axis control, created a demand for road materials which at times and at some places stretched production to its utmost limit. The companies which had the capacity and organization to fill the demand flourished mightily. Others, especially those poorly placed geographically, had a difficult time. The narratives will demonstrate the great differences between the fortunes of the various concerns during the war period.

The coming of peace produced all sorts of problems, financial, administrative, and technical. The Trunk Roads Act of 1936 had transferred 4,459 miles of principal roads to the control of the Ministry of Transport, and a 1946 Act increased the mileage to 8,190, including some roads in County Boroughs. In the same year,

[1] C. L. Mowat, *Britain between the Wars*, Methuen, 1955.

the Exchequer grants to the highway authorities were simplified, standardized according to the class of road concerned, and increased. Class I roads got 75 per cent, Class II 60 per cent, and Class III 50 per cent. The Crofter Counties of Scotland, which had been treated as a special case since the twenties, got 100 per cent. A ten-year Government programme was announced by the Labour Minister of Transport to include 800 miles of motorway. Yet in the following year, an economic crisis led to drastic restrictions on road expenditure. Indeed, it was not until 1955 that a total figure of £100m. was reached, and not until 1959 that central government expenditure exceeded that of the local authorities.

It is necessary to remind readers of the decline in purchasing power of the pound. According to one survey general building costs rose five and a half times between 1939 and 1969. The figure is doubtless lower for roads, where the labour factor is less, but it cannot be less than four, which would mean that little over twice at constant prices was being spent on the roads in 1969 than in 1939; motor vehicle population had in the same time increased at least four and a half times. Doubtless the money was being spent more effectively; certainly, productivity had much improved; nevertheless, having regard to vehicle population forecasts one may indulge in some optimism as to future demands for road materials.

To revert to the immediate post-war era, it has been established, that, at constant prices, expenditure in 1948–53 was only 66 per cent of the 1936–9 average expenditure on maintenance and minor improvement, and only 21 per cent of that on new construction and major improvement.[1]

The Government were back to the old Road Board policy of improving road crusts, and ignoring new construction. Here the technical factors became of great importance. The mechanical spreader—in these early days only the Barber-Greene finisher—had been introduced into this country by the U.S. Army during the war for the speedy surfacing of airfields, and its use on roads spread like wildfire. These years after the war showed great development in the use of bituminous toppings, notably, in the case of Tarmac

[1] *Roads, a new approach*, British Road Federation, London, p. 58.

Ltd., of 'Settite'. Both quarrying and mixing processes had greatly improved technically, and the large capacity mixer was admirably suited to the special demand. The success of the coated macadam industry in this period may be compared with the troubles of rolled asphalt, where the necessity to use large mixers, while the smaller machines were still not depreciated, led largely to the financial crises which, as our Amasco narrative reveals, were not overcome till 1962.

Despite the drop in the value of money, despite the Labour Government's rearmament programme which brought a great deal of Air Ministry work, and despite the success of 'Settite', the profits of Tarmac Ltd never reached their 1943 level again till 1950.

Though the 1910 Road Board was empowered to build roads primarily for the use of motor traffic, the first motorway in this country, the Preston By-Pass, was not started till 1956 and not completed till 1958. The Italian and German motorway experience of the twenties and thirties was ignored, and only lip-service was paid to the idea by the passing of the Special Roads Act of 1949. We may regard 1959 as the great turning-point in the attitude of the Government to road-building and road expenditure generally. As I have said, it was the first year in which central government expenditure exceeded that of the local authorities. It was the year in which M1 was opened. As regards our own special subject, it was the year in which Tarmac Ltd., embarked on a policy of acquisition and expansion which continued unchecked till the 1968 merger. The figures[1] for total road expenditure for this ten years of growth are as follows. I show the percentage increase each year alongside, but must always remind my readers of the diminishing value of money.

There is little difficulty in co-relating these figures with the financial results of Tarmac Ltd., rapidly becoming more and more a national company. We can see the main reasons for the plateau of 1959–62, the great upsurge of 1963–4, the fresh plateau of 1965–7 and the further improvement of 1967–8. It should be remembered that the expenditure statistics relate to the named year

[1] *Basic Road Statistics*, op. cit.

J. R. FITZMAURICE—CLIFFE HILL GRANITE CO. LTD.

J. N. CUTHBERT—KINGS & CO. LTD.

Year	£m.	Annual increase percentage
1959	166·0	—
1960	188·1	11·8
1961	196·2	4·3
1962	222·9	11·6
1963	266·7	19·7
1964	316·4	18·7
1965	355·5	12·4
1966	366·8	3·2
1967	405·6	10·6
1968	492·7	21·6

ending 31 March, whereas the company's financial year ends on 31 December, nine months later.

Of course, there are multitudinous minor factors affecting the comparison over sixty years of the profits of the Tarmac Derby companies with public expenditure on roads. In this brief account of demand for road materials, we have seen three times the effect of sudden demands for military purposes. Considerations of space and complexity have prevented me studying in detail the private sector of industrial roads, parking spaces and sports grounds. These have tended to vary with the general state of the economy. The nature of this chapter has excluded the effect of other demands of a non-roadmaking character, highly important in the case of limestone. Differences in managerial ability and restrictive practices legislation have moreover played their important parts.

Yet the conclusion is obvious. Increase in the prosperity of the Roadstone Division of Tarmac Derby is fundamentally wrapped up with the total expenditure on roads in this country, and in particular, with its rate of growth.

CHAPTER II

Companies Primarily Concerned with Slag

The order in which the fourteen narratives of Chapters II and III are set out requires explanation. Tarmac Ltd., as the acquiring company in all cases, comes first. Then, in Chapter II, the companies primarily concerned with slag in order of acquisition. Chapter III follows the same principle, so that Derbyshire Stone Quarries comes last.

At the end of each narrative in Chapters II, III, and IV will be found the names of those gentlemen who have furnished information and constructive criticism from their own knowledge, and without whose help this book certainly could not have been written in its present form.

TARMAC LTD. (Tarmac)

In 1901, Mr. E. Purnell Hooley, the County Surveyor of Nottinghamshire, made a discovery. A barrel of tar had accidentally burst and the tar had run over a road near the Denby Iron Works in Derbyshire. The area had been lightly covered with slag from the Denby furnaces. Hooley noticed that this portion of the roadway was dustless and resisting wear better than the adjoining stretch. He proceeded to experiment, first by tar painting and gritting and then by mixing slag with tar before it was laid. The latter appeared the better idea, and he obtained a British patent for his mixing process on 3 April 1902. He christened his new material 'Tarmac', and by mid-1903 had laid a length near Trent Bridge under particularly heavy traffic conditions. The *Newark Advertiser* of 9 July 1904 reported that this length was 'as good today as when new'.

Hooley proceeded in conjunction with Mr. John Parker, to incorporate a company called 'The Tar Macadam (Purnell Hooley's Patent) Syndicate Ltd.' on 17 June 1903. The capital was £25,000, of which £18,740 was purchase consideration to Messrs. Hooley & Parker. The working capital was found mainly by Mr. Edward Barrett. A works was erected at Denby, which finally disappeared from the scene in 1929 on the closure of the iron-works. The first mixer, designed by Ernest Miller, was built by Abbott & Co. of Newark.

Hooley himself was Chairman. A U.S. patent was obtained in 1904, and Hooley paid a protracted visit to America in the latter part of that year. However, in England things began to go awry financially largely as the result of the difficulties of the Syndicate's London agent, who was Parker's son. Some preference shares were issued, but the situation was brought under control by the inter-vention of Sir Alfred Hickman Bart., the ironmaster of Bilston. Alfred Hickman Ltd., who were becoming pressed for space to tip slag, made a slag agreement with the Syndicate, and a plant was put on order for the Springvale works at Ettingshall. Sir Alfred himself subscribed new ordinary capital in consideration of the surrender by Messrs. Hooley & Parker of a large part of their holding, and underwrote a new preference issue with the result that the Hickman interests were allotted the major part of the new capital. Sir Alfred became Chairman in January 1905 and Mr. Edward Hickman a Director in April 1905. In August 1905 the Syndicate's name was changed to Tarmac Ltd. and in September 1905, John Parker resigned. In June 1907, Hooley resigned as a Director, but was retained as an honoured consultant at a sub-stantial fee.

The London sales problem was overcome by the appointment of Mr. H. W. Fowler as London representative in February 1906. At the same time Mr. T. H. Wright was appointed General

Manager. He was, however, replaced by Mr. D. G. Comyn, the Secretary, in May 1908, Comyn being entitled Manager and Secretary.

By 1910, a business had been established. Sir Alfred Hickman died, and Edward Hickman became Chairman. The net profit in 1911 was £4,752, and Comyn reported that his orders exceeded his ability to supply and pressed the Board to secure more slag. Two hundred railway wagons were bought on H.P. terms.

In 1913 the profit was £12,396. A slag agreement was made at the Acklam ironworks at Middlesbrough, and the Board decided to liquidate the old company and transfer the assets to a new company of the same name and raise another £60,000 of capital. Tarmac Ltd., the Tarmac Derby Ltd. of today, was registered in December 1913.

The opening of World War I saw two new plants being erected, one at Acklam and the other at Brymbo, where the steel company provided the capital and the slag was to be exploited on a profit-sharing basis. Meanwhile, an old slag tip had been acquired at Bilston, not without misgivings, as the manufacture of Tarmac up to then had always been with hot freshly produced slag; Hooley's theory being that in the process of cooling, the tar oils were absorbed into the slag.

The war brought many difficulties, the profit declining from £21,792 in 1914 to £16,295 in 1918, but Comyn rapidly established good relations with Mr. (afterwards Sir Henry) Maybury of the Road Board, and obtained large orders for military purposes, moreover handing over, on a tonnage basis, the new Acklam works in 1916 to the Road Board for the supply of broken slag to the British forces in France.

As the war drew to a close, we find Tarmac taking steps for post-war expansion. Priorfields slag tip was leased from Lord Dudley, and negotiations, which proved abortive, started with Partington Steel and Iron Company in Lancashire. At the Board's first meeting after the Armistice, Comyn reviewed the position on Tees-side and asked permission to investigate the establishment of depots in the south of England for receiving crushed slag by sea, and at which Tarmac plants would be erected. He feared over-production on Tees-side. The Board in early 1919 were, however, deterred by the high cost of sea freight.

Though worried by the Tees-side prospects, Comyn was genleraly optimistic as to immediate post-war demand, having obtained a good deal of information from Sir Henry Maybury as to the Government's road programme, and was strongly backed by Edward Hickman. 1919–20 was therefore a period of almost bewildering expansion, which was financed by issue of ordinary capital which raised altogether over £300,000.

First, the policy of acquiring old slag tips in the Black Country area was expanded. Its history may be summarised as follows:

1919   Barbor's Field, Willenhall and Netherton.
1922   Botany Bay (Dawley).
1923   Friezeland tip at Millfields Road.
1924   Bilston Slag Co. (several tips) the Osier and Darlaston tips (see Crow Catchpole) and Capponfield.
1925   Stirchley, and Rough Hay and Brook Hay, near Darlaston.
1926   Hatherton.
1927   Wombridge.
1931   Mars tip, near Ettingshall.
1933   Dudley Corporation tip.

All these tips were primarily intended to augment supplies to Ettingshall works, though coating plants were erected at Netherton, Willenhall and, much later, at Coseley Road. The west Midland tip policy may be compared with that of Tarslag (q.v.).

Reverting to 1919–20 expansion, a new note was struck by the acquisition of Ffrith Roadstone Quarries on the Welsh border for £4,000. It should here be noted that apart from the small Blue Rock quarry near Oldbury and the Leigh-on-Mendip gesture of 1941, Ffrith was Tarmac's sole venture into natural stone until 1959. Ffrith itself was never developed much and disappears finally from the scene in 1951.

A third 1919 development was the acquisition of the 'Vinculum' patent, and entry into the concrete market. Though the history of Tarmac's concrete business does not enter into this book, it is to be noted that the Vinculum venture arose from a crisis as to slag dust disposal at Denby.

Before 1919 was over, geographical diversification was widespread. In North Wales, the Mostyn tarmacadam plant was leased from the Darwen and Mostyn Iron Company. In South Wales, the Cwmavon Slag Co. Ltd. had been acquired by exchange of shares (Mr. J. W. Ellis joining the Tarmac Board) and a 50/50 company called Tarmac (South Wales) Ltd. formed with Guest Keen & Nettlefolds to work the slag tips at Dowlais.

In north Staffordshire negotiations had started with the Robert Heath interests in regard to the slag at Birchenwood, and the large tip at Norton. An associated company Tarmac (Kidsgrove) Ltd. was formed in 1922, and took over the Norton tip later. Talks with Stanton Ironworks were, however, unsuccessful.

In 1920, too, an agreement was made with Palmers Shipbuilding and Iron Company at Jarrow to take the new slag, the history of which may be briefly recorded here. Palmers found the capital. In 1922 old slag was acquired at Jarrow on the adjoining land belonging to Col. Carr Ellison. In 1923 the first mutterings of quality troubles with the freshly-produced Jarrow slag appear in the Tarmac minute-book. Continuous quality difficulty followed and in 1929 the Jarrow agreement was terminated. We hear no more of Jarrow till 1964.

However, all these developments were overshadowed by happenings outside Tarmac's immediate ken. Alfred Hickman Ltd. acquired Lloyds Ironstone Co. Ltd. at Corby in March 1919, and, as a consequence, when the Prestwich (q.v.) slag agreement with Lloyds Ironstone expired in 1921, Tarmac obtained the Corby concession, and so started one of the most successful of its ventures. Stewarts & Lloyds Ltd. having acquired Alfred Hickman Ltd. in October 1920, and consequently having become substantial shareholders in Tarmac, the previous selling agents for slag at Stewarts & Lloyds' North Lincolnshire Ironworks were displaced in favour of Tarmac, who so got a start at Scunthorpe, where a plant was running by the end of July 1923.

Comyn was appointed Managing Director in September 1919, and the stage seemed set for success. Profit, subject to tax, was

£62,164 in 1920 and in 1922 was £84,261, after tax. A new name appears in the Board minutes in July 1922, when Mr. Cecil Martin, the General Works Manager, is congratulated on record outputs. Railway wagons were constantly having to be bought, and fresh capital was also required as a result of the acquisition of the Skinningrove slag on the north Yorkshire coast in 1924 and £200,000 5½% (free of income tax) cumulative preference shares were issued in September 1924. Trouble was, however, right ahead. Prices were falling, and the price of tar, partly as a result of the occupation of the Ruhr, was becoming very high. Comyn resisted pressure by his competitors to increase prices of Tarmac, and although 1,056,139 tons of tarred material were manufactured in 1924, and the profit was £83,099, after tax, in that year, it fell to £59,610 in 1925. In March 1926, Comyn resigned as a result of prolonged ill-health. A great salesman, and with an excellent head for finance, he had contributed much to the company.

So on 15 April 1926, Cecil Martin who had been appointed to the Board in November 1923, became Managing Director, a position he was to hold for thirty-two years, by which time the profits had increased some ten times. But, at the date of his appointment, the situation was most critical. The coal strike had started, and continued for many months. Jarrow was a failure. Worst of all, prices were still falling. At the Board Meeting of his appointment, Martin submitted the following schedule of prices over the last five years:

Average price of Tarmac ex—

Year	Ettingshall		Middlesbrough		Corby	
	s.	d.	s.	d.	s.	d.
1921	22	5·9	19	3·8	22	4·9
1922	18	10·9	16	8·8	21	6·0
1923	18	3·6	15	10·6	21	3·8
1924	17	5·1	13	8·4	19	2·2
1925	16	9·8	13	0·2	18	8·7

A ruthless policy of retrenchment was instituted. Directors' fees were halved, and the cost of Head Office staff reduced by 20 per cent. Meetings of the trade to improve prices proved abortive owing to non-co-operation from the steel companies marketing their own slag, Dorman Long, Stanton and Appleby. 1926 ended with a loss of £49,576, and the ordinary and the second half-year's preference dividends were passed. The support of Sir Alfred Lewis, Chief General Manager of the National Provincial Bank, was of particular value in this crisis.

The next eight years showed recovery on to a plateau of profits. The profits, after tax, were:

Year	£
1927	44,231
1928	46,570
1929	34,251
1930	42,729
1931	68,624
1932	40,671
1933	40,612
1934	42,477

They were all years of violent competition in the tarmacadam trade, but many things were happening.

Mr. C. E. Hickman, who had been appointed a Director in 1926, became Chairman in August 1929, a position he was destined to hold till 1961.

Mr. T. McMillan was engaged in 1929 to organize a Civil Engineering Department, and appointed a Director. Though the development of the Civil Engineering Division is not a part of our story, it is to be noted that it soon started to contribute to profit (notably in 1931) and that its surfacing activities, transferred to Roadstone Division in 1960, showed regular and significant margins.

It will be best to describe first the events of these eight years of national and company difficulty, during which Cecil Martin was organizing his executive control of what was becoming a big business, by reference to geographical areas, and to certain diversifications.

1. The South

Mr. Robert Hooley replaced H. W. Fowler as London representative in May 1928. In the previous eight months 20,000 tons of Tarmac had been shipped, already coated, from Middlesbrough to Gravesend, but the experiment was not successful and it was determined to revert to the policy proposed by Comyn in 1918 of setting up coating plants in the south. Other people, in particular Crow Catchpole and Silvertown Tarmacadam (q.v.) had, however, largely pre-empted the position, and it was finally decided to set up plants (opened in May 1929) at Southampton and Shoreham and to feed them with slag by rail from Skinningrove. Southampton was not a success, and the plant was moved to March in early 1932 to cope with large orders from Isle of Ely C.C. Shoreham was inevitably moved to wharfside in 1934, though its own unloading arrangements were not made till 1946–7. In 1933 a joint arrangement was made by Tarmac (South Wales) with South Western Tar Distilleries to operate a small sea-fed plant at Poole. In 1932 a joint venture was started with Constable Hart in London, which, after some false starts, took the form of a new plant at Crown Wharf, Greenwich, opened in 1933. As described in the Crow Catchpole narrative (q.v.) efforts to buy the Ford slag were unsuccessful and our period ends with Tarmac in the south interested in only three seaboard plants at Crown Wharf, Shoreham and Poole but with a great deal of Tees slag to sell.

2. Tees-side

During 1929 it was decided to set up a direct selling organization in the north-east, replacing the agency held by Thos. Swan & Co. Ltd.

In this year Dorman Long had made their great steel merger with Bolckow Vaughan and under the influence of Mr. W. F. Prentice, who later became General Manager, began to develop a positive policy of shipping slag from their South Bank works. This created competition with their tenants, Tarmac, at Acklam. Negotiations followed and in November 1932 a new deal was arranged. The Acklam lease was extended and Tarmac became selling agents for all blast furnace slag shipped by Dorman Long. Simultaneously, Tarmac concluded arrangements with Crow Catchpole and Silvertown Tarmacadam (q.v.) to sell them slag in the south, which was combined in the case of Crows with an arrangement to buy coated slag from that company. Throughout this period, Skinningrove was developing satisfactorily.

3. Derbyshire

As already noted, the first of all the Tarmac works, Denby, had to be shut down in 1929 on the closure of the ironworks. Early in 1931 a slag bank was acquired at Chesterfield and a plant opened there. Another bank, at Bestwood, was taken in 1934, but the Derbyshire problem was finally solved by a combination with Prestwich (q.v.) The Sheepbridge Iron Company gave up their own unsuccessful efforts to market their slag and gave a lease to the new company Prestwich (Sheepbridge) Ltd. in which Tarmac had a half interest.

4. Scunthorpe

Since 1923, in conjunction with Stewarts & Lloyds, the plant at North Lincolnshire Iron Company had been operative, but there were many difficulties arising from the bad state of trade both in steel and in road materials and in November 1927 the Committee of Management decided to stop the plant as the supply of slag and orders were not satisfactory. Thereafter the plant only operated intermittently until the furnaces finally went out of blast in 1931.[1] The Normanby Park Slag and Tarmacadam Co. Ltd. had, however, been acquired by Tarmac in 1924 and all the Tarmac business was transferred to Normanby, where, under various agreements with John Lysaght (a subsidiary of Guest Keen & Nettlefolds), it has flourished until this history closes.

[1] Sir F. Scopes, *The Development of Corby Works*, Stewarts & Lloyds Ltd. 1968.

5. South Wales

Cwmavon, acquired in 1919, lasted until the mid-twenties when the quality of the slag proved unsuitable. After lengthy negotiations Tarmac declined the opportunity of working the Margam slag and concentrated on Tarmac (South Wales) Ltd. in conjunction with Guest Keen. When Guest Keen erected new furnaces at Cardiff, Tarmac (South Wales) also erected a plant there and in 1931 we find them concentrated there, the capital being rearranged to take care of the losses sustained at Dowlais.

6. North Wales

A Receiver was appointed for the Brymbo Iron Company in 1931 and supplies of new slag ceased, but in 1934 Guest Keen bought the works and the business revived. Slag tips at Leeswood, near Mold were bought in 1926, and in that year negotiations were successful in renewing the lease at Mostyn.

7. North Staffordshire

North Staffordshire Slag & Tarmacadam Co. Ltd. with a tip at Fenton, had been bought in 1924. The agreement was terminated in 1931.

In 1930 the tip at Childerplay (Biddulph) had been bought by Tarmac (Kidsgrove) Ltd., and in 1934 the Heath shareholders sold out their interest in Tarmac (Kidsgrove) to Tarmac, who continued working at Norton (augmented by Childerplay) until 1938, when as described in the narrative of Derbyshire Stone Quarries (q.v.) the Five Towns Macadam Co. Ltd. took a lease of the Shelton Iron & Steel's slag at Etruria, and took over the assets at Norton. Childerplay was retained by Tarmac.

8. Lancashire

In 1928, an agreement was reached with the Darwen & Mostyn Iron Company and Tarmac erected a plant at Darwen to work the old tip.

In November 1933 Lancashire Steel determined to cease direct marketing of their tarred slag production from Irlam and Wigan. An agreement between Tarmac and Lancashire Steel for marketing the material was reached in March 1934, and so Tarmac obtained supplies of new slag in Lancashire, and a continuing business in that very important county.

9. Corby

When Stewarts & Lloyds decided in 1932 to develop Corby on the grand scale, their agreement with Tarmac still had ten years to run. It was necessary to move Tarmac, and a new agreement was negotiated for a period of forty years from 1 January 1935, and was subsequently extended to 1990. New plant therefore was put in hand at Corby.

In the meantime a Tarmac plant utilizing Corby slag was erected at Brackley in 1932 to deal with immediate large demands.

10. Diversifications

The Lilomite Asphaltic Cement patents were bought in 1927 but the process did not prove a success.

The gravel business was entered in 1929 with the purchase of pits at Bushbury and Wombourne, but profits proved hard to make. It is to be noted that the only other gravel venture prior to the acquisition of Taylors (Crook) in 1960 (q.v.) was during World War II when the Milford Gravel Pit at Brocton near Stafford was bought and then sold in 1949.

In 1928 a small Cummer asphalt plant was bought and the hot asphalt market entered, but here again the diversification was not maintained and we see Constable Hart taking a large asphalt plant off Tarmac's hands in 1932, as part of the deal in London.

During this eight-year period 1927–34 three developments of fundamental importance were taking place.

First, the normal method of delivery of Tarmac was changing from rail to road. As early as February 1924 the Board had noted that the output of Ettingshall works had been maintained during the railway strike by employing road haulage, but it was not till the beginning of 1929 that Martin convinced the Board of the necessity of buying road transport for the delivery of Tarmac, though a few Tasker 'Little Giants' with side-tipping trailers were earlier acquired to convey slag from tip to plant. A big step forward took place in the summer of 1933. Martin pointed out that the company's old Sentinel wagons would be illegal by 1940 and persuaded the Board not only to buy new Sentinels and Armstrong-Saurer diesel wagons but to get a special bank loan to finance the operation. Again in 1935 we see large buying of light lorries, culminating in 1936 in a decision to abandon the idea of erecting a rail-side plant at Brent to distribute coated Corby slag in London, but instead to buy sixty Morris Commercial light lorries and deliver direct. As a corollary of the road haulage policy, we find from 1930 onwards a continuous selling of railway wagons, which largely helped to finance it.

Secondly, bitumen was being adopted as a matrix. About 1930 Mr. H. H. Holmes of Berry Wiggins was persuading Martin of the merits of bitumen derived from petroleum refining as a binder for road material and in July 1932, Martin reported to the Board on the development of 'Settite', Tarmac's trade name for bitumen macadam. We may note here, if somewhat prematurely, that Robert Hooley who in 1937 became Sales Director was similarly reporting about 'Asphaltic Grittite', i.e. cold asphalt, in May 1938.

Thirdly, largely under Martin's leadership and encouragement, price co-operation was developing in the trade. As we saw earlier, and note from the Tarslag narrative (q.v.) Comyn had been opposed to this policy, but dire necessity in the end prevailed. Within three months of his appointment in 1926, Martin got the Board to agree to his attendance at a meeting organized by Prestwich. In December 1927, H. W. Ellis reported talks with Major of Tarslag. By early 1930, Martin reported that the movement was spreading, that a Northern Association had been formed, and that he was talking to Crow Catchpole in London. Although in 1934, the Northern Association arrangements broke down (though resumed in 1937) a London Slag Federation, with a very precise agreement was established in that year after very hard bargaining with Crow Catchpole. The London agreement was of particular value as it included contracts for the supply of Tees dry slag to the London manufacturers at substantially improved prices.

So in 1935, we find firmly established the Tarmac Roadstone Division that was to last till 1958. The producing works are there in their familiar places, Middlesbrough, Skinningrove, Scunthorpe, Corby, Ettingshall, Cardiff, Irlam, Crown Wharf and Shoreham, Sheepbridge and relations are good with four great steel groups, Dorman Long, Stewarts & Lloyds, Guest Keen, and Lancashire Steel. Natural stone plays no significant part; this is, *par excellence*, a slag business. Bitumen is being extensively used, and the road haulage department is being rapidly developed. Tarmac are leading the trade in price co-operation and are becoming very intimate with Crow Catchpole and Prestwich. Internally, Martin's management is seen everywhere. The works are well under discipline, and quality is continuously watched. The vigorous sales promotion tradition, inherited from Comyn, has been maintained and developed. Tarmac is a household name. A new stage in the story has opened.

Profits started to increase. In the old presentation of 'after tax' the company's profit up to World War II was:

Year	£
1935	60,160
1936	79,832
1937	64,975
1938	79,665
1939	57,098

In 1939, the profit, before tax, was £95,598, and my future references to profit will be in this more familiar form. It must be remembered that the company now had a civil engineering and a concrete 'Vinculum' division, the first of which was beginning to contribute significantly to profits.

New names enter the story. In 1937, Mr. A. W. Nash became Transport Manager, and after Robert Hooley left in 1942 and a period of management by Mr. H. J. Slater, Nash became Sales Manager, dying in office in 1959. Mr. L. J. Hodgkiss became Chief

Accountant in 1937, Commercial Manager in 1945, a Director in 1946, an Assistant Managing Director in 1955, and by a tragic coincidence died on 31 March 1958, one day before he was due to succeed Cecil Martin as sole Managing Director.

An interesting project of this period was a combination with Bristowes Tarvia in tarring slag at Shotts furnaces. This first excursion into Scotland, however, did not last long.

There was a big upsurge of profit during World War II, both from roadstone and civil engineering. After the initial disorganization, the airfield construction programme consumed very large quantities of coated macadam, and Tarmac were well equipped both in the size and distribution of their works to take full advantage of the position. The civil engineering contracts frequently provided a big demand. Orders, however, are reported as steadily declining in August 1944.

From a pre-tax profit in 1939 of £95,598, the wartime figures were:

Year	£
1940	143,488
1941	252,961
1942	246,786
1943	301,222
1944	265,038
1945	199,592

A most intriguing incident during the war was the negotiation to buy Roads Reconstruction, the leading limestone quarry-owners in the Mendips, with large interests in south Gloucester and Devon. Talks began in October 1940 and broke down after being within a hair's breadth of success in April 1941. To this day, Somerset, Gloucestershire and Devon and Cornwall remain the only big aggregate producing areas of Great Britain where Tarmac Derby's stake is insignificant. A small quarry was bought at Leigh-on-Mendip in 1941, but never developed. The small Blue Rock quarry near Oldbury, was also purchased as some protection against the competition of Rowley Regis (q.v.).

As the war drew to its close new agreements were entered into with Lancashire Steel, Dorman Long, and Skinningrove Iron Co. and the Board determined on an ambitious policy of works reconstruction to cope with the anticipated upsurge in demand. As a result, entirely new plants were bought into commission at Ettingshall and Acklam in 1949, Corby was completely reconstructed and a new plant was built at Scunthorpe in 1950, where a new subsidiary was formed in conjunction with Lysaghts under the name Tarmac (Scunthorpe) Ltd. In fact, in publicizing the company's jubilee in 1953, it is stated that more than £2,000,000 had been spent on reconstruction and mechanization of existing works, and the acquisition of new contracting plant, while the transport fleet had been almost entirely renewed and greatly enlarged. A minor development was into electric furnace (phosphorus) slag with Prestwich (q.v.) at Oldbury in 1951.

This stage in Tarmac's history may be regarded as lasting till 1953, when a plateau of profits was again established and lasted till 1959.

The pre-tax profits of this era were:

Year	£
1946	207,256
1947	199,420
1948	224,361
1949	291,669
1950	323,415
1951	386,036
1952	418,461
1953	680,040

A substantial contributor to the upsurge from 1950 onwards was the rearmament programme. Much work was done for the Air Ministry and the purchase of semi-mobile Black-Mix plants is to be noted.

1954 was a year of considerable significance. The troublesome negotiations with regard to the Sheepbridge slag lease were at last successful with the formation of Sheepbridge Macadams Ltd. in conjunction with Prestwich (q.v.) and Derbyshire Stone. The biggest thing, however, was Dorman Long's intimation that they proposed to construct very large new furnaces at Clay Lane (Teesport) and the decision to take the slag and build a new and massive works to deal with it, including a wharf to accommodate ships of up to 3,000 tons capacity. At first, Dorman Long were hesitant to entrust the very great marketing problem to one company, but Martin and Hodgkiss had consulted Earle of Crow Catchpole, who assured them of orders of up to 250,000 tons per annum in the south, and an offer was made and accepted to deal with the whole production. It is to be noted that much of the slag produced at the old furnaces at Clay Lane had been tipped by Dormans on the 'high tip'. The new arrangements with Dormans included the right to take this old slag. The quantity available was unknown, but was estimated at 25m. tons. Moreover, in this year, we see the replacement of petrol lorries by diesel-powered vehicles gathering momentum, and involving large capital expenditure. The Board therefore decided to make a rights issue in November 1955 of 288,605 ordinary shares at a premium of 55s. per share, raising over £1,000,000 of cash.

A great era, that of the management of Cecil Martin, was coming to its end. In January 1955 Mr. (afterwards Sir Charles) Burman and Mr. R. G. Martin were appointed to the Board, and in 1957 it was determined to turn Tarmac Ltd. into a holding company with three major subsidiaries, of which Tarmac Roadstone Ltd. was one with R. G. Martin as Managing Director. Operations under the new regime started on 1 January 1958, and it is the history of Tarmac Roadstone (afterwards to add the word 'Holdings' to its title) with which we shall, for the last ten years of this story, be principally concerned.

Tarmac's profits had got on to another plateau. The figures for 1954–8 are:

Year	£
1954	622,990
1955	747,850
1956	693,463
1957	661,256
1958	644,042

The profits were mostly derived from roadstone. Cecil Martin and Hodgkiss handed over to R. G. Martin a splendid business, but it was limited in various ways. There was practically no natural stone, and though the weakness in North Wales had been put right by the formation with Derbyshire Stone (q.v.) of Shotton Slag Ltd. in 1957 (the works opening in 1959) there were large geographical gaps. There was nothing in Scotland, nor in Cumbria and north Lancashire and there was practically nothing in all south-west England. The situation in the West Riding and north Midlands was not satisfactory. Moreover, though coated macadam surfacing was being carried on as a side-show by Tarmac Civil Engineering Ltd. (it was transferred to Tarmac Roadstone in 1960) there was no hot asphalt business though the hot asphalt market was growing. Market conditions generally looked favourable. The Government had started on a policy of heavy road expenditure, and though the passage of the Restrictive Trade Practices Act in 1956 suggested a weakening of prices, good demand and common sense were likely to overcome this difficulty. It was most noticeable that with the profitability of coated macadam, little attention had, except for railway ballast, been directed to sales of uncoated aggregate (cf. Derbyshire Stone) or to the nascent ready-mix concrete industry.

1958 saw new works at Teesport, Deptford (in partnership with Crow Catchpole (q.v.) in the Thames Tarmacadam Association) and Cardiff in production, and a new line of business entered with the first steel reclamation plant erected at Skinningrove. Further supplies of slag for Ettingshall were secured by a purchase at Lilleshall. In this year the Registrar of Restrictive Trade Practices was beginning to get moving without any immediate effect.

In October 1958 Amalgamated Roadstone Corporation (which is abbreviated to 'A.R.C.' in future references) proposed negotiations with Tarmac and Crow Catchpole (q.v.) for a merger of the three businesses. These negotiations failed, but the outcome was the acquisition of Crow Catchpole by Tarmac in August 1959.

From this moment onwards, Tarmac, and particularly Tarmac Roadstone, entered a period of great expansion, which culminated in the 1968 merger with Derbyshire Stone and Briggs. In nine years, profits increased sixfold, and the equity capital eightfold. Much of the increase in the equity was by way of scrip issues, but more represented the use of ordinary shares to acquire other businesses. Before proceeding to an account of roadstone expansion, it is worth while to consider the financial results. I therefore set out, side by side, the Tarmac Group's consolidated profit before tax and the total equity capital issued by the end of each year:

Year	Profit (£'000s)	Equity capital (£'000s)
1958	644	1153
1959	1047	3592
1960	1157	3684
1961	1243	3684
1962	1064	3798
1963	1873	5285
1964	2939	5285
1965	3343	9213
1966	3027	9228
1967	3850	9228

Roadstone normally contributed about 70 per cent of the profit, the actual figure in 1967 being £2,752,713. Broadly speaking—and this is no place for a precise financial analysis—the roadstone profits achieved a sensational increase in 1963–4, and thereafter a new high plateau was reached. Indeed in 1964, the profit of Tarmac Ltd. before taxation amounted to over 20 per cent on the total capital employed.

While major expansion was in these years by way of acquisition, we shall also relate many developments not of this nature. These necessitated large expenditures of cash, particularly on loose plant —diggers, rippers, dumpers and the like, and on road haulage vehicles. Moreover, nearly every acquisition required modernization, and therefore still more cash. When it is recalled that in 1965 the whole of Cliffe Hill and the preference capital of Hillhead Hughes were bought for cash, it is not surprising to find that by the end of the period £5m. of debentures had been issued in addition to the large increase in equity capital.

In 1959 policy was still slag-orientated, although R. G. Martin, with whom Mr. J. M. Beckett was becoming closely associated, was already perceiving that any great expansion must be in natural stone, because fresh sources of slag were not going to be easily available. Five opportunities in relation to slag, however, arose in 1959–60. First, Crow Catchpole, then in December 1959, Tarslag (q.v.) then Taylors (Crook) (q.v.) and then, through Taylor's introduction, Workington Iron Company. The only failure was at Llanwern, where Richard Thomas & Baldwins awarded the lease to competitors.

The Tarslag directors introduced Harper & Thomas, whose granite quarrying business at Bayston Hill, near Shrewsbury, was acquired, and a joint venture with Harrisons Limeworks at Shap was started to coat Harrison's limestone and Teesport slag dust.

By the autumn of 1960, therefore, great strides had been made. All the new slag in the north-east was under Tarmac's control, and they had entered the sand and gravel trade there. Tarmac, with Rowley Regis (q.v.) and Bayston Hill, had become the largest quarry-owners in the west Midlands. A start, including steel reclamation at Workington, had been made in Cumbria. Dominance had been attained in the sea-borne slag trade. An entree into hot rolled asphalt had been secured through Tarslag.

The new businesses, however, required integration, and this was no easy task, especially in the case of Tarslag. Tarslag had been organized in two geographically defined divisions, the north-east and the Midlands, while Rowley Regis, where Mr. Williams and his associates still held the preference capital, was run as a separate company. Before 1960 was over, all the roadstone assets of the recently acquired companies were transferred to Tarmac Roadstone Ltd., but preoccupation with reorganization, the undoubted success of which owed much to Beckett, who exhibited administrative abilities of a high order, probably retarded further expansion, and there were no acquisitions in 1961. A drive by Sir Charles Burman, who was appointed Chairman of Tarmac Ltd. in June 1961, for economies in overheads, stimulated administrative reorganization.

A joint venture with Dunning & Son Ltd. to operate together

Dunning's Haughmond Hill Quarry and Bayston Hill was entered into in 1961. Dover works was completely rebuilt, and major extensions finished at Shotton. Exports of slag to Germany were expanding rapidly. As a result of the Registrar's activities, the price associations had been disbanded, but new price-reporting bodies had been formed.

1962 saw the start of granulated slag exports from Scunthorpe, where orders for cement manufacture at Lübeck in the Baltic were obtained. Relations with Lancashire Steel Corporation were put on a better basis by the formation of a 50/50 company, Lancashire Roadstone Ltd. The little Filter Media Placing Company was established as a subsidiary. The drive, which we shall follow in succeeding years, to establish Tarmac as large quarry-owners in the north of England, began with the acquisition of Ord & Maddison (q.v.).

In March 1963, R. G. Martin was appointed Group Managing Director of Tarmac. Beckett and J. B. F. Earle became Joint Managing Directors of Tarmac Roadstone, Martin continuing to play a very active part as Chairman. Beckett and Earle took office after the worst ten weeks of winter weather for 200 years, with large losses on the books, but demand was so good that 1963 showed record profits.

1963 brought to fruition the Park Gate negotiations, started in 1962, by the acquisition of Prestwich (q.v.). The Park Gate Iron Co. had asked for offers for steel reclamation at their new furnaces. Tarmac, in conjunction with Prestwich, submitted an acceptable tender. Park Gate wished to add to the project the processing of basic slag, and then asked Tarmac and Prestwich to take over the old plant of T. W. Ward, who were no longer interested, to deal with their blast-furnace slag. It was clear that this old installation was quite inadequate and that a modern coated macadam works must be built. By this time the capital requirements had escalated to a point, which, as explained in our Prestwich narrative, deterred that company from continuing as partners in the venture, and the solution of a complete acquisition of Prestwich was adopted. Beckett was mainly responsible for the conduct of the very intricate negotiations with Park Gate.

The result was for Tarmac Roadstone, an assured position in the West Riding, a start in limestone in Derbyshire, a half share in the valuable Leaton quarry in Shropshire, and a big reinforcement to their sales and surfacing organization in the west and north Midlands. Moreover their new business of steel reclamation had taken a big step forward by the addition of Park Gate in Skinningrove, Workington, and Round Oak. Meanwhile Beckett's colleague, Earle, was displaying his ability to maintain good relations with the rest of the industry, which was not unnaturally apprehensive of Tarmac's very rapid expansion, especially into the new field of natural stone.

1963 saw new plant in at Middleton-in-Teesdale, Shoreham, and Etruria. The little business of Summerson at Stanhope was acquired for Middleton's protection. Inverkeithing quarry was acquired and Tarmac Roadstone were back in Scotland. Inverkeithing had been in the hands of Tilbury Dredging since 1920. It had originally been developed by the Admiralty before World War I for the construction of Rosyth Dockyard.

An interesting Board minute of 8 November 1963 reveals the belief in a continued boom, and in the necessity of finding fresh sources of aggregate. 1964 seemed to justify this optimism. With the help of exceptionally good weather, profits reached a new high record. Exports were at their maximum. A contract of the order of a million tons of slag was obtained, in conjunction with Tarmac Civil Engineering, at Teesport for the construction of the new Shell refinery, and a big bite appeared in the 'high tip' to which I have already referred.

Great attention was now being directed to hot rolled asphalt, but development was slow. Coated macadam surfacing was being pursued vigorously, not only as a business on its own account, but as a tied market, yet asphalt did not seem to get off the ground. A large new plant installed at Hailstone in 1963 had considerable success in conjunction with Tarmac Civil Engineering on the M5 and connected contracts in north Worcestershire, but a plant at Austerfield in the West Riding hung fire.

Substantial capital expenditure was undertaken in 1964 at Leyburn, Inverkeithing, Shap, and Bayston Hill. Under Martin's leadership, the acquisitions of this year included Fewsters (q.v.), New Northern Quarries (q.v.) and the old tip of good slag at Jarrow. As a result, the company's hold on slag in the north-east was completed, fresh sand and gravel resources acquired, it was set up in Cumbria and north Lancashire, and an introduction to Northumberland was effected.

1965 opened with a rearrangement of top management. Beckett became sole Managing Director with Earle as Vice-Chairman. At once Tarmac were plunged into the take-over battle at Cliffe Hill (q.v.), and by the time the year was over, Hillhead Hughes (q.v.) had also been acquired. Earle had been defeated in his attempt to buy the Pembro limestone quarries in Glamorgan and Monmouth by the pre-emptive rights of United Transport Co. Ltd. but a substantial share in them was acquired from U.T.C., who had formed a new company called U.T.C. Quarry Holdings Ltd.

Tarmac were now assured of their future in the south Midlands and north and west London, had greatly strengthened their position in Derbyshire, Lancashire, and North Wales, and improved it in South Wales and the West Riding.

However, 1965 and 1966 were hard and anxious. Government policy was acting as a brake on roadwork, and the size of the new acquisitions posed problems both of finance and efficient integration. Yet 1966 had its successes. The most important development, which does not directly concern this narrative, was the opening in conjunction with Phillips Petroleum of a bitumen refinery at Eastham, Cheshire. At last, a break-through occurred in hot asphalt with the securing of substantial orders from Durham C.C. The little business of Better Roads Ltd. was bought to speed up surfacing and asphalt development in the north Midlands, and a long lease having been obtained, the coated macadam plant at Creekmouth was rebuilt with its own unloading facilities and the first London hot asphalt unit was put in hand. In 1966 the sales of processed natural stone at last equalled those of slag.

1967 showed good improvement. Not only were the benefits of integration, of the new decentralized management structure and of the satisfactory bitumen supply arrangements becoming evident, but Tarmac were now one of the leading road asphalt concerns, even if their grip on motorway surfacing was still limited. Their plants covered Co. Durham, they were dominant in the West Midlands, and by moving the Austerfield plant to Ashton-in-Makerfield, they had entered the Lancashire market. In the south, Creekmouth and Slindon were in action.

1967 saw a new slag plant at Jarrow, and completely modernized coating arrangements at Scunthorpe, the considerable increase in capacity being promptly filled by increased orders.

Two important limestone developments had occurred. One was the purchase of the old-established Minera quarry in Denbighshire from Adam Lythgoe Ltd., which considerably strengthened Tarmac's position in North Wales and Cheshire; the other, partly flowing from the foresight of Fewsters (q.v.) the joint venture with A.R.C. at Mootlaw in Northumberland. The negotiations involving an arrangement to supply very large quantities of limestone aggregate to Ready Mixed Concrete Ltd. at their Tyneside plants, were long and arduous, but Beckett succeeded in bringing them to a very satisfactory conclusion.

The close of our sixty-five-year story is at hand. 1968 saw the erection of a plant at Tendley limestone quarry, near Cockermouth, which may be regarded as the end of Cumbrian development started before 1898 at Silverdale (*see* New Northern Quarries). Deptford works was disposed of on satisfactory terms to the Greater London Council, who wanted the site for housing. The trade was transferred to Hayes and the new large plant at Creekmouth. An important legal victory was won when the Lands Tribunal ruled that Tarmac were not liable for rates on the high tip at Teesport.

Then after protracted negotiation, came the merger with Derbyshire Stone and Wm. Briggs. Into the merger, Tarmac Roadstone Holdings brought a well-integrated business which had greatly increased in the last ten years, with very large reserves of natural stone as well as a much more complete hold on the slag industry. A big stake in the hot-rolled asphalt trade had been won by much hard work, and many of the geographical gaps, which we noted in 1958, had been filled. The fixed and loose plant, the road transport vehicles, and the commercial and sales relationships were all in good order. The market for uncoated materials was being steadily developed.[1]

CROW CATCHPOLE & CO. LTD. (Crows)

Edward Catchpole started resin distilling at Rotherhithe in 1847. Within twenty years both he and Thomas Crow were distilling tar in London. Catchpoles were in the forefront of the development of road tar as the motor age began at the turn of the century and Mr. Albert Catchpole built up excellent sales connections with highway authorities in London and the south-east. In 1920 the two small companies merged to form Crow Catchpole & Co. Ltd.

Mr. Fred Catchpole was the dominant personality and later became Chairman. In August 1923 he engaged a young graduate, J. B. F. Earle, as personal assistant. Crows' policies were largely governed by their fear of crude tar starvation. In fact, through Albert Catchpole's opportunism in getting a fourteen-year contract from the Wandsworth Gas Company in 1933, their tar and benzole distilling business lasted till 1947. However, in 1923, they determined to diversify by entering the tarmacadam industry.

They lent £25,000 to Philip Shepherd & Co. Ltd., who within a few months defaulted, leaving Crows with a large part of Bassetts Ltd., three slag contracts from ironworks in the Ruhr, and a tarmacadam plant at Rochester.

Bassetts, who held the slag lease from Shelton Iron & Steel Company at Etruria, Stoke-on-Trent, and leases of tips at Silverdale and Osier in Staffordshire did fairly well until the beginning of the thirties when unsuccessful litigation with Shelton and the exhaustion of the Silverdale tip brought the business to an end. Osier had previously been taken over by Tarmac. Incidentally, Lawford Phillips, Bassetts' active Secretary, dealt himself with the Shelton slag till the formation of Five Towns Macadam Company in 1939 (q.v. under Derbyshire Stone Quarries).

In the south-east, however, Crows' tarmacadam business began

[1] Personal sources for the above section were: J. M. Beckett; G. B. Hooley; J. M. Lindsley; Cecil Martin; R. G. Martin; Arthur Worsdall.

to grow, based on the German slag and also Tees-side supplies, originally from Wake & Co. at Port Clarence. Plants were erected at Newhaven, Deptford (after a short experiment at Barking), Littlehampton, Dover (an acquisition from Mears Bros.) and Ipswich. They also erected a tarmacadam plant at Vlaardingen near Rotterdam to supplement their coastal plants and to enter the Dutch market.

Crows were extremely cautious, and averse both from significant capital expenditure and long-term raw material contracts. The executive management of the tarmacadam business was passing into the hands of Earle, who became Managing Director in 1928, and T. C. Clinch, the Commercial Director. Crows bought raw material wherever it could be found cheaply. They switched from Germany to Belgium, where they took a share in a local tarmacadam company, and then to France. They bought Tees slag from Bolckow Vaughan, Dorman Long, and Tarmac and in the later thirties, erected small crushing plants at Ayresome and Cargo Fleet on the Tees. Crows also used much igneous rock, buying cheap parcels of whinstone from the Forth and the Tay, discoloured granite from Penlee, other Cornish granite and stone from SW. Ireland, Jersey, and Norway.

1933 saw two important developments. First, Crows obtained the contract for the slag from the new Ford furnace at Dagenham, in face of competition from Tarmac and Tarslag, and erected a plant at Creekmouth to deal with it. Secondly, offered the bait of a very cheap supply contract for limestone, they lent on debenture £20,000 to L. W. Bryant (Quarries) Ltd. of Cheddar. This opened up the west of London to them, and after brief experiments at Greenford and Wood Lane, they acquired seven acres of freehold land at Hayes with sidings alongside, and erected a fair-sized plant to exploit the limestone. They also bought slag for Hayes from a tip at Maesteg, near Bridgend, Glam.

By 1934, Crows, who had been pursuing a vigorous and successful sales policy, had established themselves as the leading coated macadam manufacturers in London and the south-east, but competition had forced prices to low levels. In association with Tarmac, they formed in 1934 the London Slag Federation and other price-fixing bodies. They obtained large shares of the various profits pools, prices were significantly increased, and Crows' profits substantially improved.

Crows now began some half-hearted efforts to secure their supplies of aggregate. They were forced to take over Cheddar quarry through Bryants' default. They acquired the freehold of Clatchard Craig quarry at Newburgh, Fife, and the lease of a small whinstone quarry at Craster, on the Northumbrian coast, and a small ragstone quarry at Ditton, Kent. Important possibilities, however, were inhibited by caution. Merger negotiations in 1937 with A.R.C. were broken off, and a later opportunity to acquire a large share in that company was declined. Though 140 acres of granite-bearing land at Markfield, Leicestershire were acquired, they were sold to A.R.C. in 1952.

World War II was a period of marking time. Sea-borne supplies were cut off. The Ayresome slag was partly marketed by Tarslag and partly diverted to a plant at York for the airfield programme. Cheddar was leased to Roads Reconstruction.

By 1948, Crows were in a curious position. Their tar distilling business had gone; they had accumulated a great deal of money by thrift and the sale of the Wandsworth tar distillery; they had rapidly re-established themselves as the leading coated manufacturers in the south-east with excellent sales connections, but apart from the Ford slag and Cheddar limestone, their aggregate situation was weak. No longer was it possible to scurry round north-west Europe buying cheap parcels of aggregate, though they negotiated a valuable contract with Mr. W. R. Northcott for sea-borne Devonshire limestone, which led to a joint venture with Silvertown Tarmacadam (q.v.) inside the main Thames Tarmacadam Association (see below), and released the site at Deptford.

This situation brought Crows and Tarmac together. Tarmac's position in the south-east was almost exactly the reverse of Crows'. Slag was pouring from the Corby and Tees furnaces, and Tarmac's organization in the south-east was not strong. Mr. Cecil Martin and Earle therefore set up the Thames Tarmacadam Association (T.T.A.), with Crows as the larger partners. By 1950, it became possible to restore price association in London with the other main producers. The T.T.A. was a very intimate joint venture, and was modified and improved from time to time by Earle and L. J. Hodgkiss, who was rapidly advancing in Tarmac's counsels. In 1955 there was a short-lived rift when negotiations for an acquisition of 50 per cent of the business by Tarmac broke down, but in 1956 a new long-term agreement was entered into, which justified Crows in authorizing substantial capital expenditure at Hayes and Ipswich. In January 1957 Mr. Fred Catchpole died and in August 1959 after the breakdown of merger negotiations between Tarmac and A.R.C., in which Crows were involved, the whole of the shares in the manufacturing and trading company were exchanged for ordinary shares in Tarmac, and Earle joined the Board of Tarmac.[1]

TARSLAG LTD. (Tarslag)

Major & Co. Ltd., tar distillers of Hull, entered into a slag agreement with the Tees Bridge Iron Co. Ltd. of Stockton in 1913. Trading as the Middlesbrough Slag Company, they erected a plant to process slag for road purposes, and took new slag from the Stockton furnaces until the latter were closed down at the end of World War I. Pease & Partners, who owned the Tees Bridge Company, made a fresh agreement with Majors in 1919 extending the lease for working the extensive tips at Stockton, and Majors then incorporated Tarslag Ltd. in 1920.

Mr. Walter Sangwin, an employee of Majors, became manager and achieved considerable success, also entering the civil engineering field. In 1922 the profits were £40,438, before depreciation of some £10,000, and Majors decided to float a new public company, and offer a large proportion of the capital to the public. This company was Tarslag (1923) Ltd.—the Tarslag Ltd. which was acquired by Tarmac in 1959. Mr. Walter Sangwin was appointed Managing Director and remained in that position until his death in 1946.

For fifteen years Tarslag made a poor showing. Profits never reached the 1922 figure and in 1929 a substantial loss was made.

[1] Personal sources for the above section were: T. C. Clinch; W. E. Witchell.

The capital was twice reduced (in 1927 and 1930). A venture into quarrying at Viroin in Belgium was unsuccessful. Mr. Major, the Chairman, and Mr. Sangwin made strenuous efforts in the twenties to bring the slag manufacturers to a price agreement with little success. The minute book is full of exasperation at the attitude of Tarmac, especially while Mr. D. G. Comyn was Managing Director of that Company. They seem to have got on better with Mr. Cecil Martin. Certainly the thirties were rather better for Tarslag, though the results show little indication of the success that began at the opening of World War II and continued until the merger in 1959.

Immediately after the 1923 flotation, the emphasis shifted from the north-east to the Midlands; the Head Office was moved to Wolverhampton in 1925. A number of old tips were acquired. Lightmoor, Corbyns Hill, and Monmore Green were entered in 1923, Stirchley in 1924, Old Park in 1925 and Pelsall in 1927. All these tips became exhausted in a few years, though Stirchley survived till 1940. The most important Midland slag venture was, however, in conjunction with Lord Dudley at Round Oak. A 50/50 company (subsequently renamed Midland Slag Co. Ltd.) started working the Round Oak tips in 1926, and the plant operated until 1941, when the slag was taken to Turner's Hill quarry (see below).

During the period up to 1939, Stockton tips continued to be worked, and the freehold was acquired in 1934. A tip at Redcar was acquired and worked out by 1936.

As the various Midland tips became exhausted, Tarslag started to move into natural stone. In 1935, a coating plant was erected at Bentley Granite Co. Ltd.'s quarry at Walsall, and operated till 1944 (see Prestwich). From 1935-40 a small coating plant was working at a quarry leased from Lord Dudley's Himley Estates at Oakham, near Dudley. This project was a failure owing to the high proportion of dirt. In 1941, Tarslag again entered the Rowley Hills by leasing a quarry at Turner's Hill from Rowley Regis Granite Quarries Ltd. and Himley Estates, to which the Round Oak slag was brought. Mixtures of stone and slag were, *inter alia*, marketed.

To conclude the story of the pre-war years, reference should be made to a venture at Dagenham, where the Essex County Council's roadstone plant was taken over in 1926 and operated on Tees slag and Belgian stone and slag till 1933, when Crow Catchpole's acquisition of the Ford slag changed the whole situation in south Essex.

In the mid-twenties Tarslag started tarmacadam surfacing by contract, and became prominent as surfacing contractors both in the north-east and the Midlands. They then entered the hot rolled asphalt market. In 1940 one of the first two Barber-Greene finishers available was bought by Tarslag.

In 1939, Mr. Douglas Sangwin, the Managing Director's son, and Mr. G. R. Baxter, joined the Board, and Mr. R. W. Oates was engaged as Secretary. The war broke out and a success story starts. Apart from an unsuccessful lime-burning project at Much Wenlock, things began to go well in all departments, but outstandingly in civil engineering. Profits (after depreciation) rose from £42,221 in 1940 to £163,595 in 1945 and remained satisfactory up to the time of the merger. On Mr. Walter Sangwin's death, Douglas Sangwin became Managing Director and remained in that position until his own early death in 1957. He pursued a policy of considerable diversification particularly into large scale civil engineering. Mr. G. R. Baxter became joint Managing Director in 1948 and Mr. S. W. Sangwin in 1955.

The roadstone side of the story divides itself, as usual, into the west Midlands and the north-east. Slag in the Midlands was confined to Round Oak where new furnaces were opened in 1951 and extended in 1958. The slag continued to be brought for processing to Turners Hill Quarry. In 1959 a small chipping plant was erected at Round Oak itself by Midland Slag Co. Ltd. However, by acquiring in 1951 the whole of the ordinary share capital of the Rowley Regis Granite Quarries Ltd. from the Williams family and their associates, Tarslag became the predominant quarry-owners in the West Midlands. The century-old history of Rowley Regis will be found in Chapter III.

Interest returned to the north-east. A limestone quarry at East Layton (Forcett) and a whinstone quarry adjoining Ord &

Maddison at Bolam were operated, but both were closed in the fifties. An asphalt plant was erected at Stockton in 1952. In 1955 a foamed slag plant was opened at Consett in conjunction with Holland Hannen and Cubitts and Mr. L. G. Clugston, Tarslag being substantial consumers of the material in the construction of 'Orlit' houses.

Main attention, however, was devoted to air-cooled slag. The quality of the remaining slag at Stockton was not good, and a right to work the Bolckow Terrace reserve at South Bank was obtained from Dorman Long. The Skippers Lane tip, which was adjacent, was also cleared. In 1955 Cleveland Slag Roads Ltd. was acquired from Thos. Swan & Co. Ltd. This provided a tip at Clay Lane, near Middlesbrough, with rights, which were disputed, to new slag from Dorman Long. There were also rights to a tip at Redcar. The Clay Lane tip was becoming exhausted by 1959, and the dispute as to new slag was amicably resolved by the formation of a joint venture with Tarmac at Acklam, and adjustments in the Dorman Long royalty arrangements. This joint venture started operating on 1 January 1959.

In 1958 the remainder of Swan's roadstone interests were acquired, comprising a plant at Consett and a lease for processing the new and old slag of the Consett Iron Co. Ltd. A large new modern coating plant was erected at Consett by November 1959.

On Douglas Sangwin's death in 1957, Mr. S. W. Sangwin and Mr. G. R. Baxter continued as Joint Managing Directors. In 1959, the Board, headed by Lt.-Com. Hans Hamilton, judged the moment ripe for merger. Tarslag were doing well, civil engineering in particular was flourishing with a most successful contract at Hartlepool. On the roadstone side Tarmac had their eyes on the aggregate reserves at Rowley Regis and Consett, and a merger was concluded in December 1959.[1]

TAYLORS (CROOK) LTD. (Taylors)

Mr. T. Taylor started the business in 1934, when he purchased from Mr. J. Proud a hand-operated sand and gravel plant at

[1] Personal sources for the above section were: G. R. Baxter; J. M. Lindsley; R. W. Oates; S. W. Sangwin.

Witton Park. He quickly modified and enlarged the plant, installing a small crusher, diesel power, and a home-made locomotive. Shortly after the outbreak of World War II in 1939, operations were suspended. After the war ended, Mr. Taylor's son, W. D. Taylor, soon became the dominant figure, taking over the management of the family company in 1951.

At Witton Park, the hoppers were enlarged, modern crushers and conveyors installed and second-hand excavators and dumpers bought or fabricated. On the sales side, W. D. Taylor obtained the sole supply of sand and gravel to the South Durham Steel & Iron Company, and established relations with large civil engineering firms. He began to build up a fleet of earth-moving equipment and obtained a private licence for open cast coal mining. In 1952 the turnover reached £149,000, and the family company was reorganized as Taylors (Crook) Ltd.

In 1954 Taylor obtained a contract from South Durham to prepare a site for an ore handling yard at Brenda Bank slag tip, near West Hartlepool. This involved handling 250,000 cubic yards of slag, and introduced him to the slag business. He erected a small crushing and screening plant at West Hartlepool for the manufacture of the rail ballast required for the ore-handling project, and was very soon supplying railway ballast to British Rail, filter media, and fines to the asphalt industry.

Meanwhile his sales of sand and gravel were increasing rapidly, and in order to conserve the high-quality Witton Park material, he bought R. B. Thompson (Thornaby) Ltd. in June 1955 in order to secure their sand and gravel resources at Thorpe Thewles, just outside Darlington. A modern screening and washing plant was quickly installed at Thorpe Thewles.

Agreement was reached in 1956 with South Durham for the whole of the new slag production at both West Hartlepool and Cargo Fleet, and work started at Cargo Fleet in 1957 on a crushing and screening plant, including ship loading facilities. This was to be followed by a coating plant, but, having engaged Mr. Lindsley as Production and Sales Manager, demand was so urgent that Taylor bought a German portable Wibau plant to tide him over until a large coating unit could be completed.

Silverdale, 1898

LIME-BURNING AND COATED MACADAM

Topley Pike, 1970

ca. 1906
LAYING BLACK-TOP
On M5—1970

Concurrent with this expansion into coated slag, Taylor started to organize a small civil engineering section, having as one object the obtaining of surfacing contracts to absorb his coated production.

Mr. Taylor's somewhat meteoric expansion was by now attracting the attention of national road material companies. At the same time, the paucity of his financial resources became increasingly obvious to him. South Durham were erecting new blast furnaces at Seaton Carew, promising a far larger slag output, and new crushing, screening and coated extensions were necessary. The coating plant was vital, because its geographical position would enable him to survive against the competition of Tarmac and Tarslag from the south bank of the Tees. He realized the impossibility of carrying on without losing independence of management and entered into negotiation with Tarmac for the disposal of the whole of the share capital of Taylors (Crook) Ltd. The deal was concluded in June 1960.[1]

WM. PRESTWICH & SONS LTD. (Prestwich)

William Prestwich had established the Gosforth Malleable Iron Foundry at Dronfield, and with it the business of a mineral and coal merchant, by 1880. He took his three sons into partnership, and their first venture into the roadstone industry was in 1898 when they started taking slag from the Denby Dale Iron Company. There followed the operation of a tip at West Hallam near Ilkeston where slag was coated with tar in a primitive plant. By 1912, development had led to the conclusion of a long-term agreement with Lloyds Ironstone Ltd. at Corby, and Prestwich dealt with the Corby Slag until 1921, when Tarmac (q.v.) obtained the rights after Alfred Hickman Ltd. had acquired Lloyds Ironstone.

New slag from the Hingley furnaces at Old Hill, Staffs., was processed from 1917 until the furnaces closed at the end of World War II. The plant continued to operate until the reserves were exhausted in the early fifties. In 1921 Prestwich reached

[1] Personal sources for the above section were: L. R. Chaplin; J. M. Lindsley.

ERM

agreement with the Renishaw Iron Company, whereby they took the slag production and, as part of the bargain, acquired from the Renishaw Iron Company the latter's limestone quarry at Intake, near Wirksworth. Though as early as 1905 Prestwich were tarring limestone at P. W. Spencer's quarry at Thornton-in-Craven in the West Riding, Intake was their first quarrying venture, and apart from the Walsall venture in 1947 (see below), Intake was the only diversification into quarrying until Berwyn Granite Quarries was formed in 1961 in conjunction with Johnston Brothers and started the very successful exploitation of Leaton Quarry in Shropshire.

Reverting to slag, the next venture was at Blaenavon in Monmouthshire. Starting in 1926, this old tip was worked (together with one at Nantyglo, entered in 1937) until its exhaustion in 1947. After a good start, the Monmouthshire ventures became relatively unsuccessful, owing to the high proportion of dirt in the tips.

Prestwich were always sales—rather than production—orientated, and consequently merchanted large quantities of slag. One source was Sheepbridge Coal and Iron Company. This ironworks was processing and endeavouring to sell its slag itself, but without much success, and in 1931, Prestwich bought the plant and obtained a slag lease. In order to extend the market, they deemed it expedient to bring Tarmac (q.v.) in as equal partners. Tarmac had practically exhausted their tip at Chesterfield and their historic hold on the Nottinghamshire market was of great value. A 50/50 company, Prestwich (Sheepbridge) Ltd. was started in 1936. In 1954, considerable difficulty arose in renewing the Sheepbridge lease. The upshot was that Derbyshire Stone (q.v.) was brought in as a third partner, and the company's name changed to Sheepbridge Macadams Ltd. A large new plant was built, but in 1961 Sheepbridge furnaces closed and operations ceased after exhaustion of the reserves.

In 1947, the end of Old Hill being in sight, the Bentley basalt quarry at Walsall, at which Tarslag (q.v.) had previously had a coating plant, was bought by Prestwich, though the reserves of available stone were not great. However, aggregate shortages in the Black Country were overcome by a novel development, that of the exploitation of the slag produced in the reduction of the mineral apatite in the manufacture of phosphorus by Albright and Wilson Ltd. of Oldbury. Negotiations started in 1947, and a contract was entered into, which was transferred to a joint venture with Tarmac which started operations in 1951. Tarmac's Production Manager had reported unfavourably on the quality of the electric-furnace slag from Oldbury, and the new source of aggregate which in fact turned out to be quite satisfactory, had slipped through Tarmac's fingers. The Oldbury slag was first processed at Walsall, but on the conclusion of the deal with Tarmac was transferred to the plant on the Birmingham–Wolverhampton road which Tarmac had installed to exploit the small Blue Rock quarry.

Prestwich's ever-active sales work had obtained for them appreciable business in the south Midlands. Before World War II, supplies had been obtained from Tarmac at Corby, but in 1950 agencies were obtained for selling slag from New Cransley and Kettering and continued till all the small south Northamptonshire furnaces finally ceased production.

An interesting interlude occurred at Intake limestone quarry from June 1959 to September 1963. Prestwich were approached by Stewarts & Lloyds, owners of the immediately adjacent Middlepeak Quarry, to take the siliceous cap of that quarry into Intake, and cease their own quarrying operations. The terms offered were so good that the deal was done, and during the period stated 4–5,000 tons per week were transferred.

The final act in the search for aggregate came in 1961. Through their Renishaw connections with Tube Investments Ltd., Prestwich in conjunction with Tarmac (q.v.) were offered the whole of the slag production of the Park Gate Iron Company's blast furnaces, together with reclamation of steel and the processing of basic slag for fertilizer from the steelworks. Though clearly a big opportunity, the problems of a family company in the third generation at once became very obvious. As will be seen from the figures set out below, the company had only emerged comparatively recently from nearly twenty years of grave difficulty, and what the family wanted, now that it was flourishing, was cash or marketable securities, not further risks. The upshot was that Tarmac bought, on satisfactory terms, all the shares in Prestwich, and by a subsidiary called Tarmac Prestwich Slag Ltd., entered in 1965 into a long-term agreement with the Park Gate Iron Company for the purposes set out above.

The following profit figures, though not properly comparative, are nevertheless illuminating. Taxation is dealt with in varying ways, and the figures are complicated from 1925 onwards by the acquisition of Wm. Lee's malleable iron foundry and other diversifications. I have averaged the figures over approximately five-year periods.[1]

	Period	Average annual profits
World War I	1915–19	£2,644 after partners' salaries and interest and taxes
Immediate post-war	1920–3	16,091 „ „ „
The middle twenties	1924–8	18,683 Net profit before tax
The depression	1929–35	2,993 „ „ „
The approach to war	1936–9	3,696 „ „ „
World War II	1940–6	1,176 „ „ „
Immediate post-war	1947–54	22,662 „ „ „
Road expansion programme		
(Stage 1)	1955–61	40,271 „ „ „
(Stage 2)	1961–5	190,173 „ „ „

SILVERTOWN TARMACADAM CO. LTD.
(Silvertown)

I have attached the narrative of this company to the slag section, in view of its early history. When Tarmac Derby acquired the majority interest from Burt, Boulton & Haywood Ltd. (B.B.H.) shortly after the 1968 merger, Silvertown was practically confined to coating limestone.

Silvertown was formed in 1925 with one-third each participation by B.B.H., the tar distillers, Walter H. Brown Ltd. the tar merchants, with whom Mr. W. H. Gatty Saunt was closely associated, and Mr. Ernest Shepherd personally, Mr. Shepherd being the

[1] Personal sources for the above section were: W. Drayton; J. Prestwich; R. W. Prestwich; C. C. Wilson.

principal of Wm. Shepherd & Sons Ltd., tarmacadam manufacturers of Rochdale.

The object was to erect tarmacadam plant at B.B.H.'s riverside
wharf at Silvertown and exploit Belgian slag introduced by Mr.
Gatty Saunt who also acted as sales agent. Mr. Ernest Shepherd,
who was regarded as the technical expert, was the first Chairman.
Mr. S. Crisp was appointed Secretary and in due course became
Manager until his retirement in 1949; he remained a Director till
1956.

The Belgian slag was mostly supplied as *tout venant*, i.e. crusher-
run, involving much granulation at Silvertown. Before long, a
good deal of difficulty arose from the 'Buy British' policy of local
authorities, and efforts were made to find suitable British slag.
This led to supplies being drawn from Dorman Long & Co. Ltd.,
starting in 1930.

Shepherd was bought out by the other shareholders in early
1930, and a difficulty having arisen over Gatty Saunt's position
now that the company were taking English slag, an arrangement
was negotiated by which B.B.H. bought Walter H. Brown & Co.'s
shares and Gatty Saunt's agreement was terminated. So from 1
January 1931, Silvertown became a wholly-owned subsidiary of
B.B.H. and remained so till 1946.

Negotiations to merge Silvertown with the tarmacadam interests
of Crow Catchpole (q.v.) broke down in 1931. Fierce competition
ensued in the London tarmacadam market. In 1934, Silvertown
joined the London Slag Federation, and prices improved significantly. There was not, however, an economic load on the plants,
and a profit-sharing arrangement with The Neuchatel Asphalte
Co. Ltd. (N.A.C.) to involve the closing down of N.A.C.'s plant
at Barking (q.v.) was arrived at. However, World War II broke
out, and in effect brought Silvertown to a standstill.

The immediate pre-war period brought other anxieties. Continuity of slag supplies from Dorman Long, whose agents were now
Tarmac (q.v.) was not assured. Small quantities of limestone and
granite were bought, but wharf conditions inhibited the successful
processing of more than one aggregate.

A small plant was operated at Mellis in Suffolk for a while

during the war. As soon as the war was over, Silvertown and N.A.C. proceeded to link more closely. N.A.C., under Mr. Vigor's leadership, became minority shareholders (finally 49 per cent) in 1946 and the post-war reconstruction of the business began. There were many difficulties. The wharf had to be dredged, and this caused much delay. Slag supplies from Tarmac were irregular and inadequate. Some ragstone could be obtained from Kent by barge, but rail supplies of any aggregate proved economically impracticable. The Board determined to concentrate on sea-borne limestone, though policy wavered from time to time, especially when a determined effort was made by Vigor to take some of the Ford slag away from Crow Catchpole (q.v.) in 1952. There was a considerable resentment at being excluded from the slag business, in which it was felt Silvertown had a well-established history, and also that they had a moral right to a supply of Dorman Long slag. Abortive efforts were made to find other regular Tees-side supplies.

Meanwhile, from 1948 onwards, regular supplies of limestone were being obtained from Moore's Quarries at Pomphlett, near Plymouth and also from Llandulas, N. Wales. Crow Catchpole were also coating Devon limestone at Deptford, obtaining supplies from Northcotts Quarries at Berry Head. Neither works was making much profit, owing to under-employment, despite price association from 1950 onwards. The time for a deal with the Thames Tarmacadam Association (T.T.A.) of Tarmac and Crows had come. By 1953 the London Tarmacadam Undertaking (L.T.U.) was set up as a partnership between T.T.A. and Silvertown. All the limestone in the east of London was coated at Silvertown, and Silvertown's slag customers were supplied from the T.T.A. works. The profits of Silvertown were much improved by this arrangement, and the friction between Silvertown and the T.T.A. members, Tarmac and Crows, disappeared. Prices were maintained by the London Coated Macadam Association of which the other original members were A.R.C. and Roads Reconstruction with George Wimpey & Co. Ltd. in friendly co-operation.

The L.T.U. worked very smoothly for over ten years. Modernization of the partners' works at Deptford, Hayes, Silvertown itself, and Creekmouth all followed and it was not till the middle sixties that its usefulness became in doubt. It was wound up in 1968.

This period was also marked by the efficient sales management of Mr. W. A. E. Rawlings. Silvertown had formed Home Counties Tarmacadam Contracting Co. Ltd. as early as 1928 with the primary object of obtaining a tied market for some of its production. Though the Home Counties Company was retained by B.B.H. when N.A.C. became shareholders in Silvertown in 1946, its orders still came to Silvertown, and with this help, Rawlings maintained sales of 80–100,000 tons per annum on the Silvertown works for many years, despite much West Country limestone competition by direct road delivery.[1]

[1] Personal sources for the above section were: A. M. Crisp; S. Crisp.

CHAPTER III

The Primarily Quarrying Companies

ROWLEY REGIS GRANITE QUARRIES LTD.
(Rowley Regis)

Igneous rocks are found in various parts of the south Staffordshire coalfield, but by far the most important mass forms the Rowley Hills, which rise to an altitude of 876 feet and measure about two miles long and one mile wide. Lying immediately south of the Black Country, being now nearly in the suburbs of Birmingham, and providing a roadstone of a high quality, the various quarries have a lengthy history which certainly runs back to the beginning of the nineteenth century.

The highest point is Turners Hill, but in the last century the outstanding feature was the basaltic mass called 'The Hailstone', consisting of a pillar abutting against a lofty slope from which it rose some sixty feet. One precise date which I can trace relating to quarrying in the Rowley Hills is that of the final removal of 'The Hailstone' in the summer of 1879. Hailstone Quarry, where Tarmac Roadstone have their local office, marks the site.

In 1900 there were twelve small quarry undertakings operating in the two square miles of the Rowley Hills. Both dressed and broken stone were produced and competition was intense. Eight of the operators joined forces under the guidance of a financier named Kirby and formed the Rowley Regis Granite Quarries Ltd. with a capital of £39,447 6% Preference shares and £37,558 Ordinary.[1] Much of the quarry land was leased from Lord Dudley.

[1] See H. W. Macrosty, *The Trust Movement in British Industry*, Longmans, Green, 1907, pp. 106–7.

Mr. W. Bassano was the first Chairman. The profit in the first year was £3,672. From the outset, one of the company's troubles was its lack of a monopoly of the Rowley stone. In a letter of 8 February 1902, we read of the breakdown of negotiations to acquire the Central Quarry of Mr. Edwin Richards. It is to be noted that this quarry never passed into the possession of the Rowley Regis Granite Quarries Ltd. during the course of this history, and was for many years the principal source of A.R.C.'s quarrying activity in the area, though R.R.G.Q. succeeded in buying the freehold from the trustees of Archdeacon Crump in 1958, when they also acquired the freehold of that half of Darby's Hill quarry which they previously leased from the Crump estate.

£14,000 5% debentures were issued in late 1902, and shortly afterwards Mr. Bassano died. Mr. Lycett became Chairman, and Mr. Simon Blewitt Managing Director. Long negotiations to acquire part of Darbys Hill quarry from executors succeeded in 1904, and things improved. In 1907 the profits were £6,849 and 6 per cent was paid on the ordinary capital.

After unsuccessful negotiations with the La Brea Asphalte Company of Birmingham, the Board decided to enter into the production of tarmacadam and a primitive plant was introduced at Darbys Hill in 1908. Two interesting documents survive from 1909. One is a spirited letter to Worcestershire County Council, accompanying a tender for tarmacadam, in which it is maintained that 'Rowley Granite is much superior to Slag.' The minutes about this time refer to the competition of the 'Tarmac Co.' of Wolverhampton. The other survival is a description by a layman of the methods of working. Though small crushers had been installed, much of the stone was hand-broken. There was considerable activity in making setts and other dressed stone. In these early years much of the stone was despatched by canal. Tracks led to the canal basins at Tividale and Hailstone and full trucks on the downward journey pulled the empties back on a funicular system. In 1910, the lease of part of Hailstone Quarry was acquired from Mr. Jones. The whole freehold was bought in 1932.

From 1907 until World War I started, the fortunes of the

company declined. Mr. Blewitt resigned in February 1914, and Mr. Mills, the Secretary, became Director and General Manager.

In 1914, Mr. T. B. Williams, son of a former Director, joined the Board, from which he retired in 1964, after having been chairman since 1943.

A local man, with both practical and commercial ability, he gave the business the attention it badly lacked. Gradually acquiring financial control, he became the dominant figure. Mr. Mills died, and when the Chairmanship passed into the hands of Mr. Bassano's son, the latter saw to it that Mr. Williams should be his successor. Mr. Williams, soon after his original appointment, arrived at a lasting commercial understanding with Mr. Edwin Richards of Central Quarry, and prices improved, while though capital expenditure was very limited, he organized the quarries in a more efficient manner. In the early twenties Mr. Williams started producing tarmacadam by modern machinery, installing a 10 tons per hour Ransome plant at Darbys Hill, but replacement of the crushers did not start till 1939 when a Broadbent plant (with a $30'' \times 16''$ crusher) was installed at Hailstone together with a Hiscox coated plant capable of 40 tons per hour. The Ransome plant at Darbys Hill was kept busy mainly by producing a fine cold asphalt called 'Regimac'.

In 1941 Tarslag (q.v.) had obtained a quarrying lease at Turners Hill. Relations were friendly and in 1951 the Tarslag offer of 26s. per share for the ordinary capital was accepted by Mr. Williams, his family and associates. A new era then started.

In the eight years before the Tarslag merger with Tarmac Rowley output rose from an annual figure of 80–100,000 tons to around 300,000 tons. Development was concentrated at Hailstone and Allsop's Hill. As soon as the acquisition was complete a road had been driven to the quarry floor at Hailstone and the services of many men and many horses rendered redundant. By March 1954 an Edgar Allen $42'' \times 30''$ crusher was working with a second Hiscox coating plant and a batch heater serving both mixers. A road was next driven into Allsop's Hill and that quarry mechanized in 1956. Under the direction of Mr. G. R. Baxter the combination of Mr. S. J. Insull on production and Mr. R. Smith on sales was bringing the old Rowley company towards the position it now enjoys.[1]

ORD & MADDISON LTD. (O. & M.)

J. R. Ord and Henry Maddison founded the business about 1850. They were both employees of the North-Eastern Railway Co., who owned a limestone quarry at Stanhope; Ord and Maddison took a lease of this quarry and started on their own. Although the first formal lease seems to be one for seven years starting 1 January 1855, they seem to have been at work for some years previously as they exhibited specimens of the stone at the Great Exhibition of 1851 and received a medal for doing so. In those early days, the major customer for the limestone was Consett Iron Co. Ltd.

The lease of Stanhope quarry appears to have been renewed from time to time until 1899. In 1863 O. & M. took the lease of Aycliffe dolomite quarry. In 1868 they started in whinstone at Middleton-in-Teesdale. John Bowes of Streatham was the landlord, and the royalty was 2d. per ton with a certain rent of £30 per annum. *The Daily Gazette* of 19 November 1869 carries an O. & M. advertisement, in which, *inter alia*, they offer carboniferous and magnesian lime and limestone, and whinstone and other road materials.

Bolam whinstone quarry was leased in 1872—the freehold was acquired in 1911—and the only other quarry in operation when Tarmac absorbed O. & M. in 1963 was Leyburn limestone quarry in the North Riding, which was taken over in 1900 from Mr. Styan, a local farmer.

The partners engaged in substantial diversifications. A patent was granted to them in September 1871 for an 'Apparatus for breaking stone' and their engineering business specialized in stone plant and agricultural machinery. A considerable range is offered in a catalogue of 1890. In 1898 Victoria Works at Darlington was acquired for the purposes of the engineering business, and for

[1] Personal sources for the above section were: Mr. and Mrs. T. B. Williams; S. J. Insull.

many years Mr. W. A. Hiscox was their manager, during which time tarmacadam paddle mixers became a speciality. In 1930, the works was sold and by 1946 the engineering business came to an end.

There were also a number of mineral ventures in coal, whinstone, sand and limestone, as far apart as Pickering and Llanbedrog in Caernarvonshire. Slag was worked at Newport, Middlesbrough from 1877 to 1901.

In 1897 the business was incorporated as a company, and we have a regular series of minute-books and annual accounts to guide us. They reveal a humdrum undertaking for the twenty-two years to the end of World War I. Average annual profits before tax were:

Period	£
1897–1901	4,353
1902–6	3,644
1907–10	4,488
1911–15	4,233
1916–19	7,588

A good deal of attention appears to have been given to Middleton. *The Engineer* of 13 December 1907 had a fully illustrated description of the plants[1] which included a 20″ × 9″ stone breaker, a cubing mill, and a paddle mixer for tarmacadam. Outputs were 32 tons of macadam and 7 tons of fines per hour, and 120 tons of tarmacadam per day. Middleton also specialized in sett-making, and there were as many as fifty sett-makers employed there at one time. The grain of the Bolam stone prevented sett-making, and it was a roadstone quarry, hand worked, until the early years of this century when a winch and a small crusher was installed. The first chipping plant was not erected till 1929. Similarly at Leyburn, which was directed to the production of fluxing stone for the Middlesbrough ironworks, there was no crushing plant till the early twenties, and the installation of a 30″ × 7″ crusher only took place in 1928.

[1] Discussed in Greenwell and Elsden, *Practical Stone Quarrying*, Crosby Lockwood, 1913.

There was a brief period of prosperity in the boom following World War I, and the profits in 1920 were £14,294. Decline, however, rapidly set in and the period 1926–34 was very bad, with frequent losses. These seem to have largely flowed from the engineering business, though Leyburn was shut for some time during the depression.

Gradually, things improved, and modernization started. In 1937 a small coating plant was installed at Leyburn. At Bolam the plant was electrified and a further crusher installed in 1938. In 1945 the adjacent Sharpley quarry was entered. By 1940, the profits were little affected by engineering activities and the next ten years show:

1940–4	average £6,560
1945–9	average £6,416

Thereafter came appreciable improvement. Better quarrying techniques and further plant at Bolam contributed, and so did mechanization at Leyburn in 1955, where the market was now nearly all in road materials. In 1957 the profits before tax reached £20,000. Though nearly £30,000 had been spent at Leyburn in 1955, further significant capital expenditure was clearly necessary, and Bolam was fully mechanized in 1960–1.

The reconstruction of the Great North Road was moving north into the area of O. & M.'s quarries, bringing about a heavy and continuing demand for roadstone in the North Riding and Co. Durham. The national companies began to show interest in O. & M. and their whinstone and limestone properties in this rapidly developing area. After spirited competition with the Tilling Group, Tarmac bought all the shares at the end of 1962.[1]

EDEN VALLEY LIMESTONE CO. LTD.
(Fewsters)

Fewsters, primarily agricultural machinery and motor engineers, opened their first sand and gravel quarry at Howford, Acomb, near Hexham in 1945.

[1] Personal sources for the above section were: A. Elliott; J. W. Elliott; G. T. Robson.

The site was only tested to a depth of about six feet, but it turned out all right, and the material proved to be of a high standard and was frequently selected by consulting engineers for concrete jobs of an exacting nature, such as the Rank Flour Mill on the Tyne. At the end of two years, the output was already at the rate of 125,000 tons per annum.

By 1961, the deposit was becoming exhausted, but the plant was kept going till 1965 by bringing in outside supplies both from land deposits and the river.

In 1956, because of the ever expanding demand for concrete aggregate, Path Head Farm at Blaydon, nearly in the suburbs of Newcastle, was bought, and a plant installed.

The Fewster management realized in the late fifties that as a result of the refusal of planning permissions, there would be a serious shortage of concrete aggregate in the Tyne valley, despite the official statements of the Planning Authorities. Since no more gravel was available, they determined to tie up Stagshaw limestone Quarry, nineteen miles from Newcastle. It is to be noted that with this quarry in their possession, Tarmac were subsequently able to arrive at agreement with A.R.C. jointly to operate the latter's Mootlaw Quarry under the style of the very successful North Tyne Limestone Co. Ltd.

Fewsters' other quarrying projects in Northumberland were the redstone (andesite) quarry at Biddlestone near Rothbury, leased in 1957, and the Whitehills whinstone quarry at Kirkwhelpington, leased in 1960. Neither was developed by Fewsters, though limited quantities of the remarkable redstone were processed at Howford.

At the end of the fifties, Fewsters determined to expand into Cumberland. They obtained the mineral rights on the very large estate of Sir Gerald Ley of Lazonby. An unsuccessful venture into mining barytes in the Hartside Fells came to an end in 1962, but in 1963 they erected a sand and gravel plant at Low Plains, as large deposits were available on the Ley estate, and the market appeared assured for some years with the programme of extension of the M6 motorway and the Manchester Water Board's new dam projects.

In 1964 Tarmac acquired all the Fewster quarrying interests.[1]

[1] Personal sources for the above section were: J. M. Beckett; K. Dixon.

NEW NORTHERN QUARRIES LTD. (N.N.Q.)

This company's predecessor, Northern Quarries Ltd. was operating at Silverdale prior to 1898. The date when it first operated Sandside Quarry is uncertain, but the paving works at Silverdale was opened in 1898, and a photograph of that date shows a lime-kiln in full operation adjacent to the paving works. Mr. James Ward was Managing Director, and writing in 1905 refers to his connection with road construction over thirty-five years.

Ward was a pioneer of tarmacadam. He produced and marketed tarred limestone under the designation of 'Quarrite', and he appears to have appreciated the importance of correct grading and of properly prepared tar in advance of his contemporaries. All the coated material was manufactured at Silverdale until 1925, when a new plant and a small tar distillery were opened at Sandside. Tar distilling had been carried on at Silverdale since shortly after the turn of the century.

By 1905, the 'Quarrite' business was well under way. A propaganda booklet of that year shows considerable stretches of paving in Blackpool and elsewhere, with testimonials from a number of local road surveyors. Northern Quarries seem to have laid much of the material themselves. It was railed from Silverdale to the Fylde, Morecambe, and Grange areas, and indeed much further afield. It seems that just prior to World War I some 20 per cent of the production was consumed in Scotland, notably in Edinburgh.

In 1909 we find Sandside producing lumpstone, some railway ballast, and burnt lime.

World War I was disastrous to the company. It was forced into liquidation, but reconstructed in 1920 as New Northern Quarries Ltd. Mr. James Ward was still Managing Director.

N.N.Q. made fair progress in the twenties. The original capital of £14,000 was increased from time to time and the dividend rose from $7\frac{1}{2}$ per cent in 1921 to 20 per cent in 1929, though 1926 was a difficult year, owing to the coal strike. Forced into buying railway wagons in 1922, as the railway would no longer carry tarmacadam in their own wagons, we can trace a gradual change to direct road deliveries, involving considerable alterations at Silverdale in 1930. By 1933 profits were being badly affected by the depression. The growth of the Quarrite business was halted, and the dividend reduced to 7½ per cent. Though N.N.Q. joined the Central Association of the Lime and Limestone Industry in 1921, chiefly through the Board's concern about rising railway rates, various overtures from its competitors to enter into price agreements failed until the 1933 troubles. Then policy changed both as regards tarmacadam and lime and in April 1934 we have a report of improved prices in Lancashire as a result of agreement with the Clitheroe producers and others. Improvements in stone production were slow and limited. For instance, compressed air for drilling was not introduced till 1930. The Board's enthusiasm was directed towards Quarrite, and they were fairly quick to accept Mr. Ward's proposals for extra tarmacadam and tar distilling plant.

In 1935 Mr. James Ward, and Mr. Wright, the Chairman, both died. Mr. M. L. Ward became Managing Director. He promptly installed diesel power at Sandside, and a certain amount of new plant including an additional tarmacadam mixer. Lengthy negotiation took place with Slag Reduction Ltd. in regard to coating the Barrow Steel slag at Sandside, but it came to nothing on the Barrow Steel Works closure in 1940. World War II brought considerable demands for ¼″ 'Quarrite' for local airfields and for limestone to the Clyde blast furnaces. The Board began consideration of modernization of Sandside quarry immediately after the war ended, but it was six years before they started a big scheme involving nearly £100,000, which was financed to a large extent by realization of investments. A new name appears in the Board's deliberations in 1942. Geo. Wimpey & Company Ltd. are reported to be erecting a tarmacadam plant at their Carnforth quarry and for twenty years, Wimpey's competition seems to have been one of the company's great anxieties. Gas nationalization had cut off much of the crude tar on which the two small distilleries had existed and the Silverdale unit was closed in 1956, and that at Sandside in 1962.

By 1960, the business was clearly in the doldrums. Mr. M. L. Ward was ill, and in fact resigned in 1961, and the Board appointed

THE ROWLEY HAILSTONE

'QUARRITE' AT BLACKPOOL, 1904–5

Mr. J. Pickard Chairman and Mr. Harold Martin General Manager. In less than a year, Mr. Martin became Managing Director, and a new era started. Modernization proceeded apace, sales were greatly extended and relations with competitors, customers, and landlords all improved. Within three years the company's position was so much better that Tarmac were enabled to make an offer for all the shares which was acceptable to the holders.[1]

The following output figures for 'Quarrite' are of interest:

Period	'Quarrite' average annual output (tons)
1921–5	26,357
1926–30	35,474
1931–5	36,392
1936–40	46,490
1941–5	37,380
(1941: 57,342)	

CLIFFE HILL GRANITE CO. LTD. (Cliffe Hill)

The crystalline igneous rock of Cliffe Hill, eight miles from Leicester, known to geologists as Markfieldite, belongs to the diorite group, and has always been regarded by road engineers as one of the most suitable 'granites' for roadmaking purposes. Prior to 1891, Cliffe Hill quarry was worked intermittently on a modest scale, but in that year it was bought by Mr. J. R. Fitzmaurice, and the business was really started. Fitzmaurice recruited Mr. Peter Preston from the neighbouring Enderby quarry and appointed him manager at £3 per week. The management passed from the first Peter Preston to his son and then to his grandson, and when Tarmac Ltd. acquired the shares in 1965, the Fitzmaurice and Preston families were still the principal shareholders. The first jaw crusher was ready to work on 2 July 1891. Preston was a stickler for quality, and rejected stone with a brown face, fining the wagon loaders involved. Cliffe Hill stone was very

[1] Personal sources for the above section were: H. Martin; I. W. Pearson.

FRM

suitable for making setts and kerbs, and dressed stone was, up to World War I, an important part of the production. Thereafter, new recruits for this arduous work were not forthcoming, output dropped by 60 per cent in five years, and by 1939 the dressed stone trade was non-existent.

In the first twelve months in 1891–2 the output of stone was 10,200 tons of broken stone and 630 tons of dressed stone. The management was enterprising, further crushers were installed, and by 1893 there were two traction engines hauling trailer wagons from the face to the crushers. The local District Council roads absorbed some of the output of broken stone, the remainder was hauled to nearby Bagworth station by tractors.

In 1894 the first big railway ballast contract was obtained (for 5,000 tons) and Fitzmaurice was encouraged to form a company. The Board decided to build a light railway to carry material to railhead and the present sidings at Bardon Hill were constructed. Work on the light railway began in September 1896, and locomotives were bought at intervals of about one year as output grew. Locomotives and rail track also replaced the tractors in the internal operations. With the increased production of crushed stone came a problem of dust disposal, so in 1903 the precast concrete business was entered. The Board started buying main line railway wagons. Preston's policy of only crushing the best grey stone was proving commercially unsound and in 1913 the crushing and marketing of sound brown-faced material was started, a subsidiary company known as 'Rockside Stone Co.' being formed to market the material. In 1913 the first tarmacadam plant (incidentally built by Ord & Maddison) was erected and located at the sidings. As an immediate consequence, the company started buying road lorries in 1914. World War I severely curtailed progress but we see considerable development soon thereafter, the quarry being electrified in 1924 and particular attention being directed to washed aggregate. We note a contract for 30,000 tons of washed filter media for Leicester Corporation being executed in eighteen months in 1926–7. In 1923 the total output, including the 'Rockside Stone' was only 80,000 tons.

Competition, particularly from slag and gravel, was getting

severe, and the Board decided to mechanize. The scheme based on a 30″ primary crusher, at that time the largest in the country, came into operation in 1929. The old 2-ton tipping tubs, hauled by a locomotive became, as a result, obsolete and after a brief and unsuccessful experiment of feeding the primary crusher by lorry, broad gauge rail, with 10-ton wagons running on it, was introduced with success.

With the introduction of bitumen as a coating material in the early thirties, another coating plant became necessary and was erected at the quarry itself. By 1937 both the coating plants had been replaced by a new plant in the quarry which operated until its replacement in 1953–6.

Loading at the face, which had been carried out by grabbing cranes since mechanization in 1929, was transferred to shovels in 1937–8.

An important event in 1936 was the purchase of the freehold of the quarries and surrounding properties from the Fitzmaurice family.

World War II brought much change. By 1949 both the founders were dead, and by 1952, the third Peter Preston was a managing director. The ready-mix concrete market was entered in 1949 and a new automatic coated macadam plant was in production by 1953. Internal haulage was reorganized by the introduction of Mack lorries; the light railway to the sidings was abandoned in favour of the company's own road transport. A new era was beginning.

From 1930 onwards, guarded references in the Directors' Minute Book indicate some co-operation with the other Leicestershire and Warwickshire quarries on prices and all the post-World War II history must be considered in the light of this. Using the long-established Leicestershire Granite Association, the quarries had a price agreement over a wide area until 1961, when they dissolved on a reference to the Court by the Registrar of Restrictive Trade Practices. Thereafter, till our story ends, limited price-reporting arrangements were made through the Federation of Midland Quarries.

To revert to the detailed story. In 1953, in order to obtain more ready-mix concrete business, Cliffe Hill acquired Davis Concrete. In 1956, a large new plant with automatic mixing of coated material was in operation and in 1958 another ready-mix plant was opened in south Leicestershire. Peter Preston was in favour of expansion by acquisition, and consideration was given to a purchase of St. Ives Gravel and others. A combination was entered into with Ready Mixed Concrete Ltd. and the Croft Granite Company and a local ready-mix concrete company, in which Cliffe Hill had a third share, was registered in 1960. Gravel pits in Norfolk were bought in 1961. In May 1963, Peter Preston died suddenly in the prime of life. Mr. J. W. Bennett, the Production Director, became Managing Director and a financial expert, Mr. Patrick Edge-Partington, was appointed Chairman. The atmosphere changed. Disposal of the family shares was in the wind. Circumstances were favourable. The extension of M1 was bringing about exceptionally good conditions for the Leicestershire granite quarries and raised the Cliffe Hill annual output to 400,000 tons. The Board introduced the shares on the Stock Exchange in 1964, acquired the local firm of Coleman Petroleum Ltd., oil product distributors, to make a bigger affair of the whole business and by February 1965, after a take-over battle with Hoveringham Gravel, Tarmac had bought all the shares for a figure approaching £3,000,000.

The profit figures over the long period 1897–1964 are well worth study, because they really only refer to Cliffe Hill Quarry activities. The management of the three generations of Preston was clearly able, and so we get a very useful picture. I have divided them, by averaging over significant periods. The presentation is of net profit *after* tax.[1]

	Period	Average net profit after tax £
The turn of the century	1897–1905	2,439
Pre-World War I	1906–14	3,725
World War I	1915–19	4,756

[1] Personal sources for the above section were: J. W. Bennett; D. Perry.

Immediate post-war	1920–3	14,248
The middle twenties	1924–8	15,202
The depression	1929–35	3,578
The approach to war	1936–9	3,829
World War II	1940–6	7,965
Immediate post-war	1947–54	11,826
Road expansion programme	1955–64	69,708

HILLHEAD HUGHES LTD. (Hillhead Hughes)

In 1921, two young men, H. Gordon Hughes and Owen M. Hughes, the sons of a professional road engineer in north Derbyshire, determined to acquire steam rollers and go into the business of hiring them out to highway authorities and contractors. Rapidly they expanded into the purchase of steam wagons, hiring them out and carrying out haulage themselves. Soon they were undertaking small road surfacing contracts in the area around Buxton, buying dry and coated stone from the neighbouring quarries. This led them to decide to acquire their own aggregate.

In 1927 they acquired the lease of Waterswallows basalt quarry on a royalty basis, but with the right, which they exercised as business developed, to pay a capital sum in lieu of royalty. Waterswallows was selected because it was an igneous rock quarry, and they were much engaged in tar spraying contracts. Before long they had set up a small pre-cast concrete works, and a coating plant. From then on, they began to specialize in coated material, using residual bitumen as their normal matrix, but taking much care in blending it, sometimes with tar.

In 1935, came the big step forward. They merged with Hillhead Quarries Ltd., owning very large reserves of limestone near Buxton and with a substantial business, and formed a public company called Hillhead Hughes Limited. Road Products Ltd., with a matrix blending and distribution business at Buxton, was also brought into the merger. It is to be noted that Hillhead Hughes never issued any fresh ordinary capital for cash. Their financial policy was conservative. Funds for expansion came from depreciation provisions, undistributed profits, and preference issues.

The Board showed much single-mindedness of purpose. They abandoned the hire business, and the surfacing. They did not persist with pre-mix concrete. They closed down lime-burning at Hillhead. Though dry limestone from Hillhead was sold in the normal markets of iron-making, sugar refining, amber glass manufacture, and railway ballast, the concentration was on coated macadam, including cold asphalt, where they triumphed over the usual technical difficulties of the use of igneous rock as an aggregate.

Their sales policy was directed to the conurbations, and south Lancashire and north Cheshire absorbed much of their production. This led to the acquisition of a hand-worked limestone quarry at Bankfield, Clitheroe in 1936 by taking over the old-established company of Richard Briggs & Sons Ltd. The quarry dates back to 1775; the Briggs company was registered in 1894. Owen Hughes was an excellent engineer and production manager. He mechanized Waterswallows in 1938-9, Hillhead in 1943, and Clitheroe in 1949-50. World War II established Gordon Hughes as one of the leaders of the coated macadam industry, and Hillhead Hughes emerged from the war as one of the important companies in the roadstone industry.

In 1952 Craig-Lelo quarry was acquired primarily to afford further supplies for the rapidly developing Cheshire market. Craig-Lelo had been opened as a hand filled quarry in 1921 and a brochure of 1930 indicates appreciable development with an emphasis on artificial stone production and tarmacadam which was first manufactured in 1925. However, World War II brought great labour difficulties and in 1951 the quarry company was forced into liquidation. Hillhead Hughes pursued a policy of modernization and development. Electricity was connected in 1953, a hydraulic flag press installed in the same year and then followed new crushers, screens, and coating plants. A conveyor system to a secondary plant some two-third of a mile long overcame many of the quarrying difficulties. Pipe-making by hand ceased, and machine-made pipes were produced on a considerable scale. In 1956 Hillhead Hughes purchased the adjoining Wern Ddu (or Dee Clwyd) quarries, which had been operating since the turn of the century, in order to increase reserves of stone. Wern Ddu was closed down until 1967 when an access road from Craig-Lelo was driven to the top quarries.

The excellent sales position of Hillhead Hughes largely flowed from the management's meticulous attention to the quality of their coated products. In furtherance of this, a central laboratory was established in 1960 at the blending and distribution establishment of Road Products at Buxton.

Further quarrying land was acquired at Clitheroe in 1961. In 1962, in order to expand in the chipping and coated macadam markets of the West Riding and Lancashire, Arcow quarry, near Settle, was bought from I.C.I. who had acquired Settle Limes Ltd. New plant was installed, and a modern coating unit was ready in 1965.

By this time, the brothers Hughes were in the late sixties, though a younger generation was growing up. Restrictive practice legislation was causing anxiety to the Board. The works which were modern in the forties had been expanded in a piece-meal fashion and required much capital for reconstruction. A long experience of mutual trust between Gordon Hughes and Cecil Martin and Earle of Tarmac created a favourable atmosphere for negotiation and an agreed merger took place in 1965, Gordon Hughes joining the Tarmac Board.[1]

KINGS & CO. LTD. (Kings)

The last quarter of the nineteenth century was not an easy one for Scottish farmers, and many were eking out a living by extracting stone from outcrops on their land, and supplying it, hand-broken, for local roadmaking. Bureaucratic difficulties were non-existent. In fact, throughout my narrative of the Scottish interests, English readers must remember that the law of Scotland is and has been much more flexible than that of England on the subject of opening quarries. It would be bewildering and irrelevant to list all the

[1] Personal sources for the above section were: T. W. Cumberbatch; H. G Hughes; O. M. Hughes.

quarries which Kings have opened in the course of their contracting history. At the same time, the ease of opening quarries, coupled with a long tradition of direct labour operations, certainly dating back to Telford (see Chapter I), has often made marketing privately quarried stone against local authorities' own production a more formidable task than south of the Border.

Robert King, a farmer of Beith in Ayrshire, was one of these farmer-quarriers. In 1899, he was in financial difficulties. Mr. J. N. Cuthbert bought his quarrying business, and with one of King's sons and two other parties registered Kings & Co. Ltd. as a private company in May 1899. The issued capital was £906, and remained at that figure till 1920.

Cuthbert at once installed a Baxter stonebreaker at Giffen Quarry, at Burnhouse, Beith. He opened two more small quarries in the Dunlop area, one at Borland in 1905, and one at Mains in 1908, where in 1909 he started manufacturing tarmacadam by a primitive process.

The first accounts I can trace are for the six months ending 31 December 1903, which show a profit of £213 after partners' salaries of £102. In 1906 'rolling' enters the picture—the first sign of contracting activity—and in 1908 the profit was £853.

We have therefore a picture of a very small concern, struggling along till World War I, but apparently already involved in small surfacing operations.

In the war, Kings diversified by entering the timber business in the north of Scotland. Sawmills were established at various sites, but as soon as the war was over, losses appeared and by 1926 all the timber business was at an end, and Kings were confined to quarrying and ancillary operations, and contracting, as they are to the end of the story. Nevertheless, they had established themselves in the north.

Records are far more complete from 1919 onwards and the quarrying and contracting department accounts are fairly well analysed. It must be borne in mind, however, that 'contracting' always included surfacing, so that the quarrying figures shown can only indicate trends, and are not of complete validity.

Average figures for profit between the wars were:

Period	Quarries and Haulage (before depreciation)	Company (after depreciation but before tax)
	£	£
1920–4	2,075	5,282
1925–9	938	1,681
1930–4	233	501 *loss*
1935–9	8,237	11,925

Briefly, these figures show the normal trends. Prosperity in the immediate post-war years; then the depression, getting worse and worse till 1933; then a recovery as we approach World War II. There are, however, many special features.

The prosperity of 1920–4 is masked to some extent by losses on timber, though general haulage with tractors was profitable. 1927 was disastrous. The first big contracting job was taken in Skye, and resulted in a heavy loss. Kings are no exception in the civil engineering industry. The results over fifty years are good, but there are three black spots—Skye in 1927, one of the hydro-electric jobs in 1948–9, and A74 at Moffat in 1960–1. These civil engineering activities are not my concern, but any reference to the company's results must be read with them in mind.

In 1927, a branch was established in the Borders, eventually with headquarters at Earlston near St. Boswell's where the company opened its first rolled asphalt plant about 1936. Asphalt development was thereafter steadily pursued. The Borders quarries, now concentrated at Craighouse and Coatsgate, have made a good contribution to Kings' profits. Occasionally the Border has been crossed. An interesting early inter-group contact took place in World War II, when Kings worked Crow Catchpole's Craster quarry in Northumberland for airfield supplies.

Many changes and development occurred in the thirties. Giffen closed, and new units in the Ayrshire and neighbouring area included Bellcraig and Middleton. Aigus quarry was opened in the north. The new units now always included coating plants. The cold asphalt market was entered with a Carpave agency. Where Kings are outstanding is in their very rapid recovery from the depression. 1938 and 1939, with annual net profits of around £23,000, when adjusted to the earlier accounting procedures, were excellent years; the prosperity is largely attributable to a combination of drainage and water schemes with a start on Royal Ordnance Factory construction.

This sort of figure was maintained through World War II. The replacement of normal work by military orders came very early to Kings, who were constructing anti-tank coastal defences at Gullane early in 1940. Later, in conjunction with George Wimpey & Co. Ltd. Kings established temporary quarrying units throughout Scotland, with many surfacing and civil engineering contracts on airfields, camps, and depots.

As soon as the war was over, an opportunity for expansion arose for Kings, which was not readily available to the other members of what is now the Tarmac Derby Roadstone Division.

The demand for electricity turned all eyes to Highland water power and for many years, the requirements of the North of Scotland Hydro-Electric Board kept Kings busy, and their quarrying activities grew at a great rate in the Highlands. Not only hard stone quarries but also sand and gravel pits were opened, and in one year nearly £10,000 was contributed to the profits from sand and gravel. This source of revenue then became insignificant until 1963.

The following average annual figures demonstrate the difference and—later on—the similarity of Kings' experience to that of their English colleagues. Too much stress must not, however, be put on the apparently overwhelming part played by the quarries. A great deal of the depreciation provision would, if the accounts had been completely separated, have been charged to the writing-off of the very large amount of new quarry plant which the nature of the business demanded.

Period	Quarries and Haulage before depreciation £	Company (after depreciation, but before tax) £
1940–4	10,291	25,140
1945–9	18,993	22,718

1950–4	24,508	19,381
1955–9	79,507	19,036
1960–4	148,074	63,849

To return to the highlights of the roadstone story. The hydro-electric programme, involving much roadwork, as in Glen Affric, required not only static plant but also mobile crushing units. A large Cedar Rapids-4 unit crushing and screening plant was imported from the United States. Much normal development was taking place. The middle fifties saw the exhaustion of the quarry at Middleton, and Craignaught Quarry, Dunlop was opened to replace it, Middleton being used as a hot asphalt mixing depot and central transport depot. By this time, much road transport was owned. Contracts, with concomitant demand for roadstone, dry and coated, were being carried out in many parts of Scotland. The new towns, industrial estates and the international airport at Prestwick provided fresh markets. The difference in demand between England and Scotland made 1956 a good year for Kings —the best till 1962, when, the troubles on A74 over, profits started a big upward trend, which, culminated (cf. Tarmac and other narratives) in a magnificent 1964. 1962 is marked by the important and spectacular Kyle of Lochalsh Road contract.

The founder, Mr. J. N. Cuthbert, died in 1946. His son Mr. J. S. Cuthbert was chief executive through these post-war years until his own untimely death in 1963, when he was succeeded by his brother Mr. W. M. Cuthbert.

1963 was a very busy year for the new Chairman. Turnover and profits were advancing rapidly. Boylestone Quarries Ltd. at Barrhead was acquired and a permanent position in sand and gravel in the industrial belt was obtained by buying the interests of Charles Tennant & Co. Ltd. at Avonside. The sand and gravel business was expanded in the last four years of this history by the acquisition of the Alexander pits in Midlothian and other pits in the south and in Easter Ross.

Then, in 1965, came the merger with Derbyshire Stone Ltd. The Group already had extensive interests in Scotland through the Amasco organization. As Chapter IV shows, Amasco were quarrying on quite a large scale in Scotland both at Croy in the west, and at various sites near Edinburgh. Our whole story of Kings shows their development as surfacing contractors and asphalt manufacturers. The three years between the Derbyshire Stone merger, and the formation of Tarmac Derby were, for Kings, a period of profitable rationalization. They took over the Amasco quarries, and promptly improved their profits by widening the marketing policy, which had previously been primarily based on supplying Amasco's own contracts. The surfacing organization and the mobile asphalt plants were similarly transferred to Amasco. Though outside our period, it is to be noted that further rationalization by Tarmac Derby Ltd. is now concentrating the management of practically all the Scottish roadstone interests of the Group under the Kings' organization in Glasgow.

It is particularly to be noted that prices of roadstone over the last fifteen years of this history had been reasonably maintained in Scotland. Reference will be made to the Central Scotland Whinstone Association later in this book, but when restrictive trade practices legislation brought much of its functions to an end, Scottish common sense maintained a sensible price level, and enabled the industry to improve and maintain standards of technical efficiency in roadwork, which twenty years previously were noticeably inferior to English practice.

Appendix: William Briggs & Sons Ltd. (Briggs)

Briggs brought various road interests into the 1968 merger, but the only parts integrated into the Roadstone Division were the quarries, now included in the Kings organization.

William Briggs started tar distilling at Arbroath in 1865, and entered the road asphalt business in 1876, when he obtained an agency for Limmer rock asphalte. The company's entry into bitumen refining was, however, delayed until 1931. Briggs were one of the earliest suppliers in Scotland of refined tar and bitumen for surface dressing, and as early as 1910 were supplying tarmacadam in Carnoustie. Between the wars a civil engineering department was formed, which had substantial surfacing activities.

Quarrying, to which of Briggs' activities this history is confined, started in 1944 with the acquisition of the Burnside Quarry Co. Ltd. owners of a whinstone quarry at Newtyle, ten miles from Dundee. In 1950 the Castlehill Sand and Gravel Co. Ltd., with its main deposits near Blairgowrie, was taken over, and finally Nairn Quarry and Brick Co. Ltd. was added. This is a red granite quarry, and, as its name implies, is in the Moray Firth area.[1]

DERBYSHIRE STONE QUARRIES LTD.
(Derbyshire Stone)

This section will be confined to the history of the limestone quarries belonging to Derbyshire Stone Ltd. in Derbyshire and north Staffordshire together with the minor quarrying activities (none of them concerned with roadstone) at Louth, Foss Cross, and Sprotborough.

It records the development of the original Derbyshire Stone business, which flowed from the merger in 1935 of the limestone quarries in Derbyshire and north Staffs of:

Constable Hart & Co. Ltd.: Cawdor and Station quarries, Matlock.

Greatorex & Son Ltd. (wholly owned in 1935 by Ragusa): quarries at Matlock and Works at Topley Pike, Buxton.

The Hartington Quarries Ltd. (wholly owned in 1935 by The Neuchatel Asphalte Company Ltd.).

John Hadfield & Sons Ltd.: Hope and Caldon Low quarries.

The Hopton Wood Stone Firms Ltd.

W. E. Constable & Co. Ltd., predecessors of Constable Hart, acquired Cawdor quarry in 1895. Station quarry was acquired from Josiah Smart much later. Mr. H. E. Giles, who played a large part in the Derbyshire limestone quarry industry in the twenty years prior to the 1935 merger had been Smart's manager. He it was who induced Mr. S. D. Clements, who had started the Ragusa Asphalte

[1] Personal sources for the above section were: W. M. Cuthbert; D. M. Young.

Paving Co. Ltd. in 1915, to buy the Greatorex quarries at Matlock from the family of that name at the end of World War I. Clements expanded Greatorex by obtaining a licence from Newton Chambers in 1932 to erect a crushing, screening, and coating plant at Topley Pike, near Buxton. Hartington had had a chequered career. Originally opened about 1911 by Norman Axe, a well-known Derbyshire character, the quarry changed hands several times before being acquired in 1928 by Wm. Shepherd & Sons of Rochdale. Shepherds went into liquidation, and after an attempted Crow Catchpole intervention, it was bought by N.A.C. in 1933.

John Hadfield & Sons, a contracting business started in 1869 at Sheffield, opened up Hope quarry from a virgin site in 1907. This quarry was finally disposed of to A.P.C.M. in 1967. In 1934 the L.M.S. Railway, who had been operating the very old Caldon Low quarry (it is believed that Brindley, the canal pioneer, who was a master millwright at Leek[1] was sometimes at work there), just within the Staffordshire border, for railway ballast production, decided to give up quarrying directly and offered the quarry to lease by tender. Hadfields were the successful tenderers—incidentally just out-bidding Tarslag.

The apparent joker in the original pack was Hopton Wood. Hopton Wood limestone from the quarries at Middleton has been well known from the eighteenth century as the most attractive limestone for architectural and monumental sculptural work in the country. Epstein and Eric Gill were only two of twentieth-century sculptors who employed it extensively. Another sombre activity was the supply of 120,000 headstones to the Imperial War Graves Commission after the close of World War I. The two original quarrying companies, Hopton Wood and Killer Brothers, had merged in 1905 and control was in the hands of the Salmon family. The quarries in other hands offered potential dangerous competition to the other Derbyshire producers.

At the beginning of 1935, the Derbyshire limestone producers found themselves in a sorry plight. The combined profits of the five undertakings which were to succeed in merging to form Derbyshire Stone Ltd. had fallen very sharply as set out below

[1] Smiles, *Lives of the Engineers*, John Murray, 1861.

side by side with the total road expenditure in Great Britain for the nearest corresponding year:

Year	D.S. constituents' profits (before depreciation) £	Total road expenditure in G.B. £'000
1930	68,726	66,373
1931	77,585	68,787
1932	62,762	53,953
1933	40,195	50,756
1934	44,349	51,501

Lack of demand had its usual effect on prices. The Derbyshire Limestone Producers Association had been of benefit in securing a measure of restriction upon competition and price-cutting, but it was felt that the arrangements were unsatisfactory, since Members were free to leave the Association at will and since the penalties for over-selling were insufficient.

John Hadfield took the initiative in bringing together the five undertakings whose earlier history we have been tracing. Apart from the overwhelming motive of reducing competition, the five companies were unanimous in looking for unified management in order to concentrate production at the larger and better equipped quarries, and to reduce selling and distribution expenses.

Hadfield's action was a courageous one. The idea of a merger of five parties might well deter anyone, and there was a special difficulty in the fact that Hadfields had only just acquired the Caldon Low quarry from the L.M.S. Railway and could show no past figures of profit there. The crucial meetings after some months of negotiation took place on the 16 and 30 September. Salmon of Hopton Wood showed considerable skill in mediation between the parties, and agreement was reached on 14 October 1935.

Merchant bankers had been consulted and had expressed the view that the shares could not be immediately placed having regard to the unsettled international situation and the stock market's prejudice against purely quarrying concerns, so the parties agreed to finance the merger on their own. This included finding £60,000 working capital.

They decided that the initial capital should be £426,000 divided equally between 5½% Cumulative Preference Shares and Ordinary Shares.

Derbyshire Stone Limited started business on 1 January 1936 with headquarters in Dale Road, Matlock, at what had been Greatorex' office. Constable Hart, who had brought a large number of railway wagons into the merger, were the largest shareholders, and Major Owen Hart was appointed Chairman. The new Board's wisest appointment, however, was of John Hadfield as Managing Director. At the age of 42, he got his chance.

As we study the history of the next thirty-two years, we must continually recall that Derbyshire limestone enters many fields far removed from roadstone, and one important part of Hadfield's great contribution to what is now Tarmac Derby's Roadstone Division is his continuous application to these other markets.

The first acquisition Derbyshire Stone Limited made was in 1936 when Geo. Lovegrove & Co. Ltd. operating Hoegrange quarry near Ashbourne was bought for £4,000 to remove a potential source of competition and closed down. In 1937 Ashwood Dale near Buxton was bought with the same motive from Highways Construction. This quarry of peculiarly high quality limestone is now chiefly employed in the manufacture of industrial powders. Similarly competition from Bradwell Dale quarry near Sheffield was eliminated.

1938 saw the passing of the Agricultural Act and the inception of the Land Fertility (Agricultural Lime) Scheme. Derbyshire Stone acted swiftly, introducing grinding mills to produce the appropriate grades of ground limestone and rapidly reached a leading position in the market.

A spectacular event in July 1938 was the first 'Big Blast' at Caldon Low. 100,000 tons of stone was shattered, and much publicity gained with the help of Lord Stamp, Chairman of the L.M.S. Railway, who detonated the charge by pressing a button in his office at Euston.

In 1938, too, the company started buying its own lorries, financing the acquisition by selling railway wagons.

World War II brought many problems, but an interesting

acquisition was that of the Taddington basalt quarry in 1944 from Ben Bennett Junr. Ltd. In 1940, the fluorspar business was entered and, due to the enterprise of Mr. J. W. Hobday, the Production Director, before long Derbyshire Stone became the largest producers and processers of metallurgical fluorspar in the country. The extraction and processing of fluorspar ended in 1963 on the conclusion of an agreement with Laportes. The Roadstone Division still markets most of the metallurgical fluorspar produced by Laportes.

Mr. S. D. Clements, who had succeeded Major Owen Hart as Chairman of Derbyshire Stone, died in 1945 and John Hadfield became Chairman as well as Managing Director. This fresh appointment, and the end of World War II, introduced a new era of conspicuous growth and development.

The age of mechanization and automation was beginning. All the works required rebuilding and modernization. Fresh capital was required and £300,000 of $4\frac{1}{4}\%$ Cumulative Preference Shares were issued in 1946.

1948 and 1949 were difficult years on the roadstone side, but the production of raw materials for agriculture, iron and steel, and other industries was at a high level.

A distinguishing feature throughout the Company's history was the consideration given by the management to the welfare of the work people. Relations with labour were good, and in 1949 the Company pioneered the first workmen's non-contributory pension scheme in the quarrying industry in the country. Later a non-contributory life insurance scheme was introduced. The safety of the employees received special attention. In 1964 Hadfield declared publicly that he would rather see the Company's profit adversely affected to the extent of £100,000 than have any employee killed while engaged in the company's service.

1951 saw a yet larger 'Big Blast' at Caldon Low. This time 278,000 tons of limestone were dislodged. The agricultural demand caused Hadfield to look beyond Derbyshire, and an oolitic limestone quarry at Foss Cross in the Cotswolds and a chalk quarry at Louth in Lincolnshire were acquired. The latter, normally producing some 120 tons per day, came into great prominence in February

GRM

1953 when the abnormally high tide and gale force winds destroyed the sea walls of the Lincolnshire coast and flooded the towns of Mablethorpe and Sutton-on-Sea. Extreme urgency was required to rebuild the sea walls, and, by a considerable feat of organization, due chiefly to Hobday, the output of the Louth quarry was increased to 4,000 tons per day for the period of the emergency.

Despite the limited demand for roadstone in the early fifties, the Company's profits continued to increase. The slag industry had been entered in 1939 when, in conjunction with Tarmac, the Five Towns Macadam Company Ltd. had been established at John Summers' Shelton Iron & Steel Company furnaces at Etruria near Stoke-on-Trent. After the war, Summers built new blast furnaces at Shotton in Flintshire, and Derbyshire Stone acquired an option on the slag in 1952. This led to the formation of Shotton Slag Ltd. in 1957, with Tarmac as equal partners, and the works was opened in 1959. There had been a long period of negotiation, arising out of initial difficulties as to the quality of the slag at Shotton, and out of Tarmac's fears as to the effect on their Irlam venture in conjunction with Lancashire Steel, but the rapidly increasing demand for roadstone soon made these apprehensions a memory. The third slag interest was at Sheepbridge, which I describe under the history of Prestwich.

Derbyshire Stone also entered the Leicestershire granite industry by acquiring Forest Rock quarry near Coalville in 1948. Development proved difficult, and the quarry was sold on satisfactory terms to a consortium of the other Leicestershire granite producers in 1960.

Hadfield was always in the forefront of co-operation in the roadstone industry, and the history of Derbyshire Stone Quarries must be read with this in mind. The very formation of Derbyshire Stone was his first great achievement in this direction. In 1936 we find him presiding at the inaugural meeting of the British Limestone (Roadstone) Federation. In 1951 he became the Chairman of the Federation of Coated Macadam Industries, which he had visualized in his inaugural speech of 1936 and which had been formed in 1944. Apart from these technical and propaganda bodies, however, he was always ready to talk to his competitors on commercial affairs, and when the price-fixing associations in the roadstone industry were challenged by the Restrictive Trade Practices Act of 1956, we find him Chairman of the Committee of Roadstone Producers who organized the industry's defence of its practices on a national basis.

To return to the narrative of Derbyshire Stone Quarries. 1954 saw the end of the getting of Hopton Wood stone in block form for architectural and similar usages. Costs had become prohibitive with excessive overburden. The remarkably pure limestone with its calcium carbonate content of over 99 per cent and the extraordinarily low iron content of only 0·02 per cent is just what the glass-making industry requires, so Hopton Wood activities were concentrated in the unique Middleton limestone mine and the business flourishes in altogether different markets.

I have referred to Topley Pike in my account of Greatorex' history. In 1955, Newton Chambers ceased quarrying and leased the quarry to Derbyshire Stone who started extensive modernization schemes in this large and geographically strategic quarry in 1962 which came to full fruition in 1964. This included the erection of lime kilns for supplying lime to the metallurgical industries, as well as a large coated macadam unit. Wredon in Staffordshire was acquired in the same year.

In 1961, the two Harry Evans companies were acquired, specialists in the spreading of agricultural limestone and in the transport by tanker of limestone and similar powders. In this year, Mr. Harold Fletcher became Managing Director of Derbyshire Stone Ltd. Derbyshire Stone Quarries Ltd. had been formed for administrative and taxation reasons in 1959. When Mr. Fletcher retired in 1963, Mr. George Henderson became Chairman and Managing Director of the quarrying subsidiary.

In 1963, as a result of the acquisition of the controlling interest in Amasco, the Sprotborough magnesian limestone quarry near Doncaster, came into the quarrying group. It had been originally acquired by John Hadfield & Sons in 1927.

The final major acquisition in Derbyshire limestone came in 1966. Middlepeak quarry near Wirksworth was privately owned and worked by hand till 1948 when it was acquired by Stewarts &

Lloyds Ltd. mainly for the supply of metallurgical limestone and lime to Corby. They decided to mechanize and the work was carried out in the period 1953–6. The siliceous cap on the quarry was unsuitable for metallurgical work, but excellent for roadstone, and this stone was for over four years supplied to Prestwich (q.v.) at the adjoining Intake quarry. The necessity of building new and larger coating plants and the prospective shortage of limestone reserves at Matlock (due to planning difficulties) coupled with the imminence of steel renationalization created a situation which was favourable to negotiation and Derbyshire Stone acquired the Middlepeak quarry (often known as Bowne & Shaw from the original owners) for £1·7m. with a long-term contract to supply metallurgical limestone and lime to the Stewarts & Lloyds works at Corby and elsewhere.

On 31 December 1966, Mr. John Hadfield retired from the office of Executive Chairman of Derbyshire Stone Ltd. During the thirty-one years that he had been managing the business, the profits had risen from some £40,000 to £2·6m. He was succeeded by Sir Edward Senior.[1]

[1] Personal sources for the above section were: J. S. Collinge; H. Fletcher; John Hadfield; G. Henderson.

CHAPTER IV

The Story of Amasco

Amasco (Amalgamated Asphalte Companies Ltd.) was a merger in 1957 of The Neuchatel Asphalte Co. Ltd.'s U.K. asphalt interests with Highways Construction Ltd. and practically all of Ragusa Ltd. Ragusa had previously acquired Chittenden & Simmons Ltd., and in 1962 Amasco acquired H. V. Smith & Co. Ltd. I have consequently decided to set out the narratives of the five constituents in order of their formation dates, and conclude with a narrative of Amasco itself from 1957 to 1968.

THE NEUCHATEL ASPHALTE CO. LTD. (N.A.C.)

The history of this company concerns us in a limited way. It was registered in 1873 for the purposes of acquiring the concessions granted by the canton of Neuchatel to mine rock asphalte in the Val de Travers and to take over a number of companies which were marketing the asphalte in various European and South American countries. It did not, however, market the asphalte directly in the United Kingdom, because it had assumed an obligation to grant the exclusive right to operate here to a totally separate undertaking, Val de Travers Asphalte Paving Co. Ltd. to which it supplied the asphalte. This right was terminated by arrangement in 1922, and N.A.C. entered the U.K. market directly, its interests extending in due course beyond asphalt to coated macadam and quarrying. In 1957 N.A.C. became a participant in the Amasco merger, and its direct roadstone interests in U.K.

were limited to a quarry in the Mendips, disposed of in 1960, and a sand and gravel pit in Kent, which it transferred to the Roadstone Division on the conclusion of the Tarmac Derby merger. A small amount of asphalte rock is still bought by Amasco's road division for the manufacture of mastic asphalt.

1873–1922

It was in 1712 that a Greek, Dr. d'Eyrinis, discovered a bed of natural rock asphalte—a band of very pure soft limestone ten to fifteen feet thick impregnated with about 8 per cent of bitumen— outcropping in the Travers valley. In 1873, at the instigation of certain Geneva bankers, N.A.C. was formed for the purposes already described.

As the nineteenth century drew to a close, the increase of traffic demanded better city streets, compressed Rock Asphalte found great favour and the demand spread rapidly for both rock and asphalte powder from cities all over the world.

After a shaky start, profits which in 1873–4 were only £866, by 1900, reached nearly £33,000. From the outset, the Board were grumbling about the excessive royalty payable to the cantonal government, and this developed into a running fight, with the Grand Council of the Canton starting an action in the Swiss Courts, but there was a friendly settlement and a new concession running to 1925, and on more favourable terms, was granted in 1896. The latest concession extends to 1991, 118 years after the first one was granted to N.A.C. In 1894 fresh mines were purchased in the Abruzzi range in Italy (but disposed of later) and many years later at Seyssel in France.

The company was originally greatly over-capitalized at a total of £1,137,000 divided into ordinary and preferred shares. A major dispute arose between the two classes of shareholders which culminated in the action, famous in accounting history, of *Lee* v. *N.A.C.*, which ended in a decision in the Company's favour by the House of Lords. The action sought to prevent the directors paying dividends before making provision for the 'wasting asset' of the mine. Arising from this, a capital reconstruction was sanctioned by 'The Neuchatel Asphalte Company Ltd. Act of 1892'. The capital was reduced to £419,880, all in one class. We note thereafter issues of new Preference Shares in 1900, 1901, and 1908. The original stock market quotation was obtained in 1883.

From 1900 to 1914 N.A.C. continued to prosper, profits reaching nearly £50,000 in 1913. World War I upset everything, and a great deal was lost as a result of devaluation of foreign currencies. The ordinary dividend was passed for some time. Yet by 1922, the company was back again in its stride with annual profits around £40,000.

1923–1945

Having been released from its obligations to the Val de Travers Asphalte Paving Co. Ltd., N.A.C. now determined to enter the U.K. market directly. In 1924 the Board reported that: 'During the past year the necessary arrangements for carrying out Home contracts have been completed, including the installation of plant at the Company's Works, Nacovia Wharf, Fulham. A good start has been made in executing work in London and the Provinces.' Optimism in regard to the U.K. soon began to evaporate, and in 1926 and 1927 the branch showed heavy losses. The directors held a private meeting on 29 September 1927 and decided to instruct chartered accountants to review the position. In considering their report dated 24 April 1928, it must be remembered that it refers to asphalt activities both in road-making and building. It criticized the Board severely for entrusting the General Manager, who died in 1924, with a General Power of Attorney. It was suggested that the general lay-out of the wharf was wrong, that wrong and even useless plant had been bought, and that a disastrous contract had been entered into for Italian bitumen. At any rate, of the £261,000 invested in the English business by the end of 1927 over £100,000 had been lost. These losses were masked by good results from abroad, but the situation became apparent soon afterwards as the following statement of Company profits between the wars will show.

Year	Profit	Year	Profit	Year	Profit
1921	£39,932	1927	£51,545	1933	£11,134
1922	38,532	1928	45,278	1934	24,177
1923	43,371	1929	31,612	1935	30,526
1924	62,382	1930	23,570	1936	32,912
1925	54,916	1931	15,850	1937	66,438
1926	46,212	1932	10,412	1938	43,108

The Board were slow to act on this report, but by 1931 we find Mr. William Cash, Senior, in the Chair, and Mr. A. E. Nicholson, F.C.A., Managing Director. In 1933, Mr. Eric Vigor, who had been with the company since he was a very young man, became manager of the U.K. business. A new era of management had started, and it was quickly reflected in the profits, though U.K. business remained highly competitive, but turnover was growing. The company entered the coated macadam business with a plant at Barking based on sea-borne slag. This was brought into a joint venture with Silvertown Tarmacadam Co. Ltd. (q.v.) and after World War II the interests were consolidated by N.A.C. becoming minority shareholders in an enlarged Silvertown Tarmacadam Co. N.A.C. entered quarrying with the purchase of Hartington Quarries in 1933, soon to be exchanged for shares in Derbyshire Stone in 1935. Terry Quarries in Kent were acquired in 1936, the Cray River gravel business in 1937, Mendip Stone at Fairy Cave quarry near Frome in 1940 (disposed of in 1967), Croy and Ratho quarries in Scotland in 1946 and 1955 respectively.

The start of World War II was destructive to road demand in the U.K. but by 1941, demand for airfields and other military works began to fill the gap. Mr. Vigor returned from war service in 1945, and was immediately appointed General Manager. In 1950, after the death of Mr. Nicholson, he became Managing Director.

1946–1957

During World War II, the American Air Force had introduced Barber-Greene finishers for the laying of bituminous materials and wet-mix on airfields. When the war was over, there was dissension in the asphalt industry as to the application of mechanical finishers to roads. N.A.C. were pioneers in their use, and gained a temporary advantage over their competitors. The use of the Barber-Greene machine rapidly led to the acquisition of asphalt plant of much greater capacity. Previously 20 tons per hour was the maximum size. Larger contracts were being let and N.A.C. took their full share of these, sometimes in remote places, such as Belfast. The original contract ended, and a nearly-new asphalt plant of considerable capacity remained. Fresh business was sought, and so we see the establishment of large branches, such as in Northern Ireland. The trouble was that the smaller plants were not worn out and total capacity soon became far in excess of demand. Profits, which had been rising, began to decline, as the expansion of the road programme continued to be delayed. Hadfield of Ragusa then approached N.A.C. with a view to merger with N.A.C.'s U.K. interest. N.A.C.'s reaction was favourable but they thought that Highways Construction should be asked to participate. The Cowdray Group had management problems and were willing to place the management of the merged concerns in the hands of Vigor. So Amasco was born. It had far too much plant, largely rapidly becoming obsolete, and we shall find that its early years were very difficult. The final point to make in this short narrative of N.A.C. is that it was N.A.C. management, headed by Vigor, which had the task of making Amasco a success.[1]

CHITTENDEN & SIMMONS LTD. (C. & S.)

Mr. E. B. Chittenden founded this Kentish business in the eighties of last century by entering the market for the ownership and hiring of agricultural machinery, in particular threshers. This rapidly led him and his partner Mr. P. A. Simmons into traction engines and steam rollers, and they carried and rolled considerable quantities of Kentish ragstone. The next move was into ragstone quarrying and when the business was incorporated in 1906, we find them in possession of ragstone quarries at Offham, Borough Green, and Allington, and stone and gravel quarries at

[1] Personal sources for the above section were: G. A. Gould; E. Vigor.

Sevenoaks. The Chittenden family held 1,803 shares of £10 each out of the total capital of 2,325 shares. Mr. Harry Moor was the Secretary.

C. & S. were very early entrants into tarmacadam and tar paving, and rapidly built up a reputation for the supply and laying of these materials in all the south-east counties, London, and by utilizing sea transport, as far up the East Coast as Yarmouth. Mr. R. G. Chittenden, at the age of 20, was already taking a hand in the business in 1909. Mr. E. B. Chittenden died in 1913, and Mr. R. G. Chittenden became Managing Director. One of his first activities was to enter into an agreement with the Trinidad Lake Asphalte Co. for the purchase of an asphalt plant and its erection at Borough Green and for the supply of Trinidad Asphalte. In 1913–14 this asphalt plant must have been one of the first asphalt units erected for supply to rural roads. World War I started; R. G. Chittenden joined the Forces and in early 1915 Mr. Simmons died. Mr. Harry Moor stepped into the breach and successfully managed the business through the war years. On R. G. Chittenden's return from service, he threw himself with great energy into development, both on the quarrying and asphalt and surfacing sides. The agreement with Trinidad terminated and he formed a company with Messrs. Wettern and Beadle called Associated Asphalt Co. Ltd. This became a success, and was operated under a very friendly agreement whereby C. & S carried on its own asphalt business which was mainly confined to Kent and East Sussex.

Chittenden was becoming an important figure in the road materials industry. In 1928 C. & S. became one of the parties to the large quarrying merger of British Quarrying Co. Ltd. of which the dominant figure was Sir Henry Maybury. Borough Green and Allington ragstone quarries passed into this merger, and are now an important part of the business of A.R.C. Mr. Moor joined B.Q.C. as Kent Manager.

The asphalt and surfacing business was retained, and continued to flourish. By the time World War II was over, Mr. N. F. Machin was playing a considerable part in management and by 1949 he was Managing Director of the operating company.

A number of schemes for purchase or amalgamation were

considered, but finally an offer from Ragusa, whose activities ran parallel, seemed the most favourable, and C. & S, was sold to them, passing eventually into the Amasco merger in 1957.[1]

H. V. SMITH & CO. LTD. (H. V. Smith)

A company under this name was formed in 1909 to acquire a tar spraying business being carried on by Mr. H. V. Smith and Mr. H. D. Martin, who was a qualified civil engineer and played the leading part throughout the early history of the company. Mr. Richard Winch acquired shares in 1910 and soon bought out the original Smith holding, which had passed into the hands of Mr. F. D. Johnson. The other substantial shareholder was Sir John Oakley, who was chairman for many years till he was succeeded by Mr. Stanley Winch, Mr. Richard Winch's son, during World War II. Mr. E. North was the Secretary from the start. The Company was reasonably successful and in the latter years of World War I began to make good profits. In 1919 it was liquidated and a new company of the same name formed to take over the assets.

Having obtained a limited concession for the sale of Trinidad Asphalte, a 25 tons per hour asphalt plant was erected at Tottenham prior to the war, and in 1920 another was opened at Dagenham. Mr. Sampson's name first appears in 1922. He was later transferred to Scotland where he managed a branch, with a depot at Inverkeithing, till his death in 1948.

The post-war era was very profitable and in each of the three years 1923–5 a dividend of 25 per cent was paid on the capital which gradually increased from £41,885 to £51,777. Mr. Humphrey Martin, Mr. H. D. Martin's son, was appointed a Director in December 1925. Though specializing in tar spraying, the asphalt and surfacing departments were developed, and a fair position in the industry began to be established. From 1926 to 1931 things began to get more and more difficult; the dividend was being reduced until it was only 7½ per cent in 1931. 1932–4 were critical,

[1] Personal sources for the above section were: R. G. Chittenden; N. F. Machin.

losses being incurred, but from 1935 onwards till the end of World War II, under the management of Humphrey Martin, North, and Mr. Miller, who had been appointed Directors in December 1930, considerable improvement took place. Again, however, the company got into difficulties, and there was no dividend from 1945–52. Scotland was a particular anxiety, and when Sampson died in 1948, his successor Mr. Douglas McPherson was under strict instructions to confine his activities to tar spraying and tarmacadam surfacing in the summer months.

1952 brought dissension on the Board, and the upshot was the appointment of North as Managing Director. Though things then improved with a resumption of dividends in 1953, Humphrey Martin and Mr. David Winch, grandson of Mr. Richard Winch, who had joined the Company in 1948, were dissatisfied. North fell ill, and finally ceased to act as Managing Director in August 1957, after nearly fifty years of devoted service.

In 1959 Martin and Winch were appointed respectively Chairman and Managing Director. Winch began to apply modern methods of management. The dividend from 1957 onwards was at the rate of 15 per cent. Civil engineering, which had been disastrous, was closed down in 1961. McPherson joined the Board, and the Scottish business prospered; an asphalt plant was installed at Kaimes quarry of the Corstorphine Quarry Co. Ltd., afterwards acquired by Amasco in 1963.

David Winch began to acquire much respect in the industry, and Amasco acquired the company, and Winch's services, in the summer of 1962.[1]

HIGHWAYS CONSTRUCTION LTD
(Highways Construction)

The group headed by Lord Cowdray formed this company in 1913, with the original purpose of developing the disposal of the bitumen being produced as a residue by the Mexican Eagle Oil Company, which was owned by Lord Cowdray until a controlling

[1] Personal sources for the above section were: D. C. B. Winch; D. McPherson.

interest was sold to the Royal Dutch Shell Group in 1919. In 1923 Highways Construction began to market in the United Kingdom the rock asphalte coming from the French Asphalte Company's mining concession at St. Jean de Maruéjols near Avignon, which had originally been granted to a French consortium in 1859, had been acquired by British interests in 1906, and by the Cowdray Group in 1923.

For forty years Sir Henry Goldney was Executive Chairman of Highways Construction, while the manager was Mr. H. F. Berry, almost from its inception till his death in 1947. The Highways Construction management seems to have been in the forefront of the development of hot rolled asphalt utilizing residual bitumen, which after 1919 was normally bought from Shell. Business was very good in the twenties and expansion was rapid, with substantial branches in Scotland, the West Riding and Lancashire. A later speciality was 'rock-non-skid', incorporating French asphalte rock shipped via Cette (Sète).

As hot rolled asphalt developed, the Highways Board determined to acquire sources of aggregate. Ashwood Dale limestone quarry near Buxton in Derbyshire was one project, but a more interesting venture was that into slag. Slag interests were acquired from 1928 onwards. The Oakengates and Donnington tips were bought from the liquidator of Wm. Shepherd & Sons and then Wake & Co. Ltd. was taken over in order to control the tip at Port Clarence on the Tees. A tip at Thornaby was also taken. It is to be noted that no contracts for new slag were ever made.

In 1934 the London coated slag manufacturers formed a consortium under the leadership of Tarmac and Crow Catchpole to control the distribution of business and to raise prices from the uneconomic level to which they had fallen. Tarmac were faced with the problem of disposing of substantial quantities of new slag from Teeside and Corby in the London Market. A scheme was devised whereby Highways undertook to ship no slag to the Thames in consideration of approximately 15 per cent of the profits of the London Slag Federation. The scheme was successful and the benefit to Highways substantial. Prices were significantly increased and by rationalization of deliveries from the various plants in London, transport economies were effected. The Federation continued until 1940, when it was disrupted by World War II. As a result, no slag was shipped by Highways from the Tees tips. Indeed, little slag was extracted from them. The Thornaby quality was indifferent, and though some material was utilized in West Riding contracts, the stone content of road asphalt was usually igneous rock. Policy seems to have changed. Ashwood Dale quarry was sold to Derbyshire Stone in 1937, and little exploitation of the slag in Shropshire, which again was of indifferent quality, took place.

After the war, fresh problems arose for Highways. Freight rates from Cette became prohibitive. Mr. Berry's death raised management difficulties for the Cowdray Group, and the development of the Barber-Greene finisher and consequent proliferation of asphalt plants raised the same difficulties which I have described in the story of N.A.C. The formation of Amasco in 1957, in which Lord Poole played an active part, was therefore welcome.[1]

RAGUSA LTD. (Ragusa)

When Ragusa merged with Highways Construction and the U.K. interests of N.A.C. to form Amasco in 1957, the Ragusa Asphalte Paving Company's business was practically confined to asphalt for building purposes. Ragusa's road-making subsidiaries, however, are today integrated in the Roadstone Division of Tarmac Derby. These were Chittenden & Simmons, Tarpaving and Tarmacadam Ltd., and John Hadfield & Sons.

Ragusa was formed by Mr. S. D. Clements in 1915. We have already traced (see Derbyshire Stone Quarries) the history of Ragusa's quarrying ventures in Derbyshire through the acquisition of Greatorex. Ragusa never seem to have mined asphalte in Sicily, though they owned some asphalte rock bearing land in that island. Their active natural rock asphalte interest was at Montrottier near Annecy, though little rock was exported to this country;

[1] Personal sources for the above section were: Sir T. Blomefield, Bart.; F. W. Fry; H. W. Watson.

there was a big trade, through Cette, with North Africa. The interest was sold in the early fifties.

Tarpaving and Tarmacadam Ltd. was started as a wholly-owned subsidiary in 1925 to carry out surfacing work of all kinds. Under the management first of Mr. Holmes who specialized in school work and then of Mr. Cantrell, it developed a substantial business. After the Amasco merger, the business was integrated in the new company.

Mr. Clements died in 1945. At first profits increased from the level of the war years, but in 1951 they were at a very low ebb, and the Company was further embarrassed financially by undertaking the construction of a large new mastic asphalt factory at Matlock.

Institutional shareholders became concerned, and Mr. Ian Henderson was appointed as Chairman. In his opinion the building side required the support of road asphalt activities and Ragusa therefore acquired Chittenden & Simmons.

I have described the history of this very old-established Kentish company separately. Thereafter Mr. John Hadfield entered the picture. He sold John Hadfield & Sons, old-established paving contractors of Sheffield and Birmingham to Ragusa for shares, and, joining the Board, soon became Chairman.

The end of the story was Mr. Hadfield's promotion of the Amasco merger which absorbed nearly all Ragusa's assets.[1]

AMALGAMATED ASPHALTE COMPANIES LTD.
(Amasco)

The grave difficulties which had brought the three parties to the merger together did not vanish overnight, and it was in fact four years before Amasco was on a profit-earning basis. On 19 November 1962, just five years after the date of the merger, the Court sanctioned a reduction of capital of £755,000 which more or less represented the accumulation of losses to the end of 1961.

Amasco had a very experienced Board, including Sir William

[1] Personal sources for the above section were: J. Hadfield; J. A. A. Spring; F. W. Fry; R. J. Cantrell.

Cash, Lord Poole, and Mr. John Hadfield, and they supported their Managing Director, Mr. E. Vigor with loyalty and persistence in his difficult task through this very lengthy period of frustration.

Market conditions in all departments were poor and notably so in the roads division. The Air Ministry demand of the early fifties, which we have already noted in the narrative of Tarmac (q.v.) had come to an end, and the very large hot rolled asphalt requirements of the motorways had not yet developed. The outstanding competitor was Limmer & Trinidad, excellently organized and by no means prepared to concede to Amasco that share of the trade which at the start only the size of the new company could justify. In fact, what can only be described as a price war raged for some three years, while Vigor was carrying out the arduous task of integration and welding the three (or rather five, for Chittenden & Simmons and John Hadfield & Sons had separate organizations) components of Amasco into an efficient unity.

Though the financial resources of the three shareholders were very large, they were by no means readily available. Capital expenditure was limited to inevitable replacements, with one exception.

This was the creation of one works in London, to replace the five being operated by the constituent companies. The site chosen was at Pear Tree Wharf, Greenwich. It belonged to the Cowdray Group, and was let by them to Amasco, who proceeded to erect asphalt units for both road and building purposes, and a jetty to handle vessels of up to 1,800 tons.

Meanwhile, efficiency was being gradually brought about in the other parts of the United Kingdom. Staff economies on a large scale were introduced, individual depots which existed in many of the major cities were concentrated, old plant was scrapped. The necessary writing off of many items of course contributed to the accumulating loss.

One promising development which arose in 1960 was regretfully abandoned mainly for lack of funds. Mr. R. Preston-Hills, who as Assistant Managing Director, supplied a great deal of support to Vigor throughout this trying period, had succeeded in obtaining an option from Taylors (Crook) Ltd. (q.v.) for 100,000 tons per annum of sea-borne Tees slag from Cargo Fleet. The coated macadam industry in London was flourishing in contradistinction to hot asphalt, and it was proposed to utilize the wharf and site at Greenwich to build a large plant and enter into competition with Tarmac (q.v.) in the coated slag market.

1962 brought success at last. From 1 January, all the business was conducted in the name of Amasco. The market improved, and so did Amasco's share of it, their competitors being now convinced that it could not be treated merely as an enlarged Neuchatel Asphalte Company. Prices rose, and with a greater turnover profits commensurate with the reduced capital were earned. It had been found, after the first wave of economies of integration, that little more could be done in the direction of reduction of overheads.

In the improved climate, the shareholders provided £500,000 by way of unsecured loan stock, and Amasco was therefore able to make some acquisitions. In July 1962, the entire share capital of H. V. Smith & Co. Ltd. (q.v.) was bought, and Mr. D. C. B. Winch joined the Board, and became an additional Assistant Managing Director.

The quarries in Scotland, leased from N.A.C. (q.v.) had proved a satisfactory investment, and £270,000 was spent in modernizing Croy in 1963-4. In September 1963, Corstorphine Quarry Co. Ltd. was purchased. This brought the total production of the Company's Scottish quarries up to 400,000 tons per annum, of which at least 50 per cent was coated material.

In September 1963, a fundamental change took place. Derbyshire Stone and N.A.C. bought out Ragusa and the Cowdray Group, with the result that Derbyshire Stone owned 60 per cent and N.A.C. 40 per cent of Amasco. With Derbyshire Stone's acquisition of N.A.C. in 1965, Amasco became a wholly-owned subsidiary of D.S.

Despite the appalling weather at the beginning of the year (cf. Tarmac) 1963 results were good, and for the year ending 31 March 1965, they were excellent, consolidated net profit before tax amounting to £539,246. In July 1965, Vigor retired as Managing Director in favour of Winch, but remained as Chairman.

HRM

In 1965 Derbyshire Stone also acquired Kings & Co. Ltd. (q.v.) Rationalization followed, with Amasco's Scottish quarries being transferred to Kings, and Kings asphalt and surfacing activities being transferred to Amasco. The Amasco profits, therefore, are not wholly comparative in the last three years of our story, but may be described as remaining on a high plateau, with a good advance in the year ending 31 March 1968.

The roads division made a big contribution to the success of these three years. Though much motorway surfacing was undertaken, this market did not prove a good one from the profit angle, an experience shared throughout the industry. It was the upsurge in spending in the cities which provided comfortable margins, and Amasco was well placed to take advantage of this, especially in London.[1]

[1] Personal sources for the above section were: E. Vigor; D. C. B. Winch; L. H. C. Mills.

CHAPTER V

Slag—and Iron

In districts where iron was being worked the hard slag provided
an almost ideal metalling.

I. D. Margary: *Roman Roads in Britain*, I. 1955.

The charcoal iron furnaces of the Romans and of our less remote
predecessors are now only of romantic interest, though it may be
observed that from medieval times the Sheffield and Birmingham
areas were the centres of iron-using cottage industry. To understand the modern slag industry in which the Tarmac Derby
companies have played such a predominant part, we must however, look briefly at the history of the modern iron industry, not
least at the motivation behind its geographical location. It would
be foolish to pretend that before the end of World War I, the
fortunes of iron-making had any significant effect on the infant
slag industry; the quantities of slag utilized were far too small.
With the 1919–20 expansion of Tarmac narrated in Chapter II,
however, started that intimate connection which explains much
of the narratives therein.

Modern iron-making starts with the successful experiments
with coke-furnaces by Abraham Darby at Coalbrookdale in
Shropshire in 1709. Production of iron in the eighteenth century
was, however, very small, and there were significant imports of
Swedish and Russian charcoal iron.[1] The situation changed

[1] Birch, *The Economic History of the British Iron and Steel Industry 1784–
1879*, Cass, 1967, p. 19.

towards the turn of the century. Cort's invention of the puddling process for making wrought iron,[1] coupled with the armament demands of the Napoleonic Wars and the threat of loss of naval control of the Baltic, created minor boom conditions, markedly in the Black Country with its iron-using crafts, and in South Wales.[2]

The slump that followed Waterloo continued till about 1830, when again technological advance and a sudden new market combined to prosper the iron industry, with ups and down, for at least forty years. Neilson's invention of the hot-blast process not only reduced fuel costs sensationally, but enabled the coals and ores of the Glasgow area to be utilized, bringing Scottish iron very much into the picture.[3] The railway era created a demand for rails and other iron goods which lasted long after the British railway building era, as exports poured into the developing railways overseas.

Already we are in contact with our subject. Between 1830 and 1852 the output of the Black Country furnaces rose from an annual figure of 211,604 tons to 725,000 tons.[4] A House of Commons Report of 1849 lists 118 iron furnaces in blast in south Staffordshire at fifty-one sites.[5] I have identified fourteen of these sites in the names of the tips worked by our companies, as set out in Chapter II. The fifties were, however, the zenith for the Black Country iron-making industry. The famous 10-yard seam of the local coalfield became exhausted, and so did local ore. By 1879 only forty-four furnaces were in blast, and iron production was halved.[6] Today only the Hickman furnaces at Ettingshall remain, and the slag reserves have vanished.

The second half of the nineteenth century brought great changes in the British iron industry, and as a consequence, established the slag industry in the locations where it has flourished from 1920 until now, with two exceptions, Corby, of which more later, and the Ford furnace at Dagenham.

First, the developed railway system made it generally more advantageous to carry coke to ore than vice versa.[7]

[1] Birch, op. cit., pp. 33 ff. [2] Op. cit., p. 45. [3] Op. cit., pp. 26 ff.
[4] Op. cit., p. 133. [5] Op. cit., p. 390. [6] Op. cit., p. 156.
[7] Op. cit., p. 331.

Secondly, vast new orefields were developed. Vaughan found the outcrop of the main Cleveland ore-bed at Eston, near Middlesbrough in 1850; in the same year, Schneider discovered the great bed of hematite ore at Barrow.[1] In 1859 came the discovery of the reserves of north Lincolnshire, and at the same period the realization of the value of the huge quantities in Northamptonshire.

Thirdly, Bessemer's invention of the pneumatic process for making steel, first announced in 1856, initiated the age of cheap steel. Siemens' open hearth process followed, and there was an enormous boost to iron making. Steel production increased from 1·3m. tons in 1880 to 7·7m. tons in 1913.[2]

However, neither of the original Bessemer and open-hearth processes could function with phosphoric ore; the only native hematite ore was in Furness and Cumberland. The fourth great factor was therefore the basic open-hearth process of Gilchrist and Thomas operating on phosphoric ore. This was only slowly accepted by ironmasters, except on Tees-side, where large new works on the principle opened in 1880.[3]

I can only describe the consequence of these changes as they affect slag in the broadest way. The detail is fascinating and sometimes relevant as at Ettingshall, where Sir Alfred Hickman's utilization of the great reserves in the Black Country of phosphoric tap cinder in the Gilchrist and Thomas process[4] partly explains the survival of that ironworks when all the other blast-furnaces in south Staffordshire have disappeared.

First in importance to Tarmac Derby comes Tees-side. The first blast furnace was blown-in in 1851; by 1871 there were 122 furnaces and the Cleveland ore production was nearly 5m. tons. To appreciate the social consequences of this very rapid development I recommend my readers to read the masterly chapter on Middlesbrough in Professor Asa Briggs' *Victorian Cities*. The Tyneside and north-west Durham furnaces all finally closed with

[1] Carr and Taplin, *History of the British Steel Industry*, Blackwell, 1962, p. 12.
[2] Andrews and Brunner, *Capital Development in Steel*, Blackwell, 1951.
[3] Carr and Taplin, op. cit., p. 102.
[4] Carr and Taplin, p. 102.

the exception of Consett, leaving behind slag tips, which, in the case of Jarrow, are not yet exhausted.

Secondly, South Wales. The great ironmasters of the valleys, notably Guests at Dowlais, were forced to the ports in order to get hematite ore from Cumberland and Spain by sea cheaply. Iron-making was transferred in 1891 to East Moors, Cardiff, though the other activities at Dowlais continued. In the 1849 Report[1] thirty-six furnaces were in blast at Cwmavon, Dowlais, Maesteg, Blaenavon, and Nantyglo, all of which places figure in the narratives of Chapter II. Today only Ebbw Vale remains as an iron town in the valleys, largely for sociological reasons.

Thirdly, north Lincolnshire. The ore was found to be self-fluxing, and this economy, coupled with the proximity of the West Riding market, brought about a development at Scunthorpe second only to Middlesbrough. It may be noted that Tarmac were effectively late on the scene, owing to the failure of the North Lincolnshire Iron Company, and have never dominated the Scunthorpe slag industry in the way they have on Tees-side.

Fourthly, the north Midlands. Here the situation and reaction to the new conditions was complex and spread over a very long period. A great factor was the presence of a huge iron- and steel-using market in the West Riding and the west Midlands. A great disadvantage was distance from the ports, as foreign ore, first because of its non-phosphoric properties and then because of its richness in iron, became of ever-increasing importance. For instance, the building of the railway from the Gellivare district of Sweden to Narvik brought a new fillip to Middlesbrough in the early years of this century. The Derbyshire furnaces, however, had the benefit of the near-by Northamptonshire ore-field. The upshot was concentration into a few very large ironworks, such as Stanton, Park Gate, and Shelton at Stoke-on-Trent. We may include in this paragraph the iron industry of Lancashire and North Wales, now concentrated at Irlam and Shotton, both accessible to foreign ore.

I will not take up space in relation to Scotland and Cumbria. Apart from the minor venture at Shotts, the Group has never been

[1] P. 98, above.

in the Scottish slag industry. The hematite slag of the Cumbrian ironworks is practically useless as aggregate, and the great heaps at Barrow, Cleator Moor, Maryport, etc., bear witness to the fact.

It is difficult for us to enter into the political mentality of the great Victorians, including such outstanding ironmasters as Sir Lowthian and Sir Hugh Bell of Middlesbrough. They were passionate Gladstonian free-traders. As the twentieth century opened, foreign iron and steel was pouring into this country dumped under the protection of foreign tariffs, but the political atmosphere was still violently anti-tariff reform. The British iron trade was fairly prosperous, because of the world-wide demand for steel, but relatively it was rapidly losing ground to the continental and United States producers. Then came World War I, which disorganized the unprepared British iron trade, and so we come to 1919, when, as I have said above, the fortunes of the iron trade really began to matter to the Tarmac Derby companies.

The steel trade contemplated the end of the war with high hopes. They believed that there would be a great demand to replace the ravages of war, and to carry out all manner of capital schemes inevitably postponed during the four years of conflict; they thought too that Germany was knocked out. At first, their hopes were realized. The price of Cleveland pig iron, which was £7. 2s. 6d. per ton in May 1919 rose to £11. 5s. 0d. by January 1921. Then came the slump. By April 1922 the price was £4. 10s. 0d.; in January 1926 it was £3. 9s. 0d. Depreciated currency, territorial changes causing the dissolution of trade associations, large supplies of battlefield scrap, and workers accustomed to low living standards under blockade conditions, all contributed to violent competition, against which the unprotected British industry was helpless.[1] Carr and Taplin list eleven leading British steel companies who in 1921 paid their last ordinary dividend for ten years.[2]

By this time, blast-furnace slag had established itself as a first-class aggregate for road-making, in particular in the coated form. Purnell Hooley had proved to be right; slag, owing to its porosity had a better affinity for binder than natural stone,

[1] Carr and Taplin, op. cit., pp. 361 ff.
[2] Carr and Taplin, p. 365.

especially than granite. At the same time, it was stronger than limestone and did not polish easily. Another factor was giving slag a commercial advantage over its competitors, and that was the geographical location of ironworks. Either they were, as a result of iron history, in areas where there was no first-class natural aggregate, as at Scunthorpe and Irlam, or they were in the midst of conurbations, as at Renishaw, Cardiff, Tees-side, and north and south Staffordshire. The industry was, however, beginning to experience the effect of being an unimportant byproduct of a major industry. In any particular area, there is always too much or too little good slag. Here the quarry-owners' great economic advantage lies. If trade is bad, reduce quarrying; if it is good, open another face. These options are not open to those operating on freshly-produced slag; nor, indeed, in modern capital-intensive quarries; while tips of good old slag last, however, slag producers can escape from the crises brought about by the closing of furnaces.

Here I must pause to refer to the subject of slag quality. Many of the slag tips created in the nineteenth century and in the first two decades of the twentieth were very unsuitable for road material purposes. Hot-blast slag was suspect; because of some old slag made by this process being useless, there was a good deal of prejudice against freshly produced slag generally. Sometimes specifications demanded 'old cold-blast slag'. A genuine trouble was excess of silicon, producing a vitreous material to which tar would not adhere. This was markedly the case in South Wales— and why Tarmac abandoned Cwmavon. Slag tips were often so contaminated with ash and other rubbish that the cost of extraction of good slag was prohibitive. See, for instance, Prestwich's troubles in Monmouthshire.

So far as freshly-produced slag is concerned, ironmasters refused to concern themselves about quality. They found enough technical trouble in solving their iron-making problems. In 1866 Kitson told the Coal Commission that no West Riding ironworks manager understood the simple elements of chemistry,[1] and the build up of scientific management from that time was a very long and difficult task. Slag is a mixture of complex minerals. Until 1942 when the

[1] Birch, op. cit., p. 11.

British Standards Institution published the emergency wartime B.S. 1047 'Slag for concrete manufacture', following the invaluable researches of Dr. T. W. Parker of the Building Research Station, there was no scientific definition of good slag. This was the turning-point. The blast furnace manager now at least knew what was wanted, and the unique characteristic of slag, i.e. that its nature can be changed while in the molten state, made it possible to produce stable slag under what had been previously regarded as unfavourable conditions. Ironmasters began to move away from the attitude towards slag epitomized in the introduction in 1934 of the author to some American engineers by the Chief Engineer of the Ford Motor Co. Ltd. 'Meet', he said, 'my gentle laxative!'

To return to the period of the great iron and steel depression. Staffordshire, both north and south, became an area of too little new slag. So we see the tips of the Black Country, the Potteries and nearby Shropshire seized and their content of good slag exhausted, while transport considerations caused a contraction of the economic market, and London and the south-east was largely abandoned to sea-borne slag. In the north-east, with Tees-side iron surviving every catastrophe, there was too much slag, and prices remained very weak. Tarslag remained content with their great tip at Stockton (only finally exhausted in 1969). The eyes of Dorman Long and Tarmac turned to the sea-borne market, only to find that the Germans and the Belgians had got there first, and that Crow Catchpole and Silvertown (q.v.) and others had snapped up the best coastal sites. In Derbyshire, though Renishaw survived, Denby closed, and Tarmac and Prestwich were very glad of the Sheepbridge opportunity. There was a Receiver in at Brymbo, and Tarmac's position in North Wales was weak for twenty years. North Lincolnshire Iron Co. came to an end, and Tarmac were driven back to Normanby Park.

At last came the turning of the tide for the British steel industry, and, albeit very sluggishly at times, it has been flowing ever since. In 1931, the output of pig iron was lower than it had been since 1859.[1] In September 1931, Britain was forced to leave the gold standard. By February 1932 the Import Duties Act was law. The

[1] Carr and Taplin, op. cit., p. 472.

Act provided *inter alia* for an Import Duties Advisory Committee, and the steel industry came under a measure of Government control, which in the end led to nationalization.

Coincident with the start of recovery in the steel trade came a development of the greatest importance to Tarmac. Despite the great ore-field, iron-making in Northamptonshire, which started in 1852 at Wellingborough, had only been on a small scale. Lloyds Ironstone were making iron at Corby when, through Alfred Hickman Ltd., they passed into the hands of Stewarts & Lloyds in 1920, and as our narratives describe, the comparatively small production of slag was allotted to Tarmac. In 1932, however, Stewarts & Lloyds decided to develop Corby on the grand scale,[1] and Tarmac entered into possession of a very large source of new slag of the highest quality, moreover situated nearer to London than any other first-class road-making aggregate, except for the production of the Ford furnace at Dagenham, and Kentish ragstone.

In these years just preceding World War II, a slag development of a technical nature was also taking place. Road engineers had long been looking for a cheap substitute for much of the hot rolled asphalt they were consuming, and research and development into cold asphalt, largely associated with the name of Dr. Dammann, had been going on for some time. W. F. Rees found the commercial solution in bituminized slag dust which he christened 'Resmat', but it was the slag entrepreneurs, notably Tarmac, who had the problem of disposing of this by-product of a by-product, who reaped the major benefit. Slag was much preferred to granite for this purpose, because of its tolerance and affinity for binder, and the cold asphalt trade for the last thirty years has been very largely in slag, having reached a demand of some $1\frac{1}{4}$m. tons per annum. It is noteworthy that pre-1940 slag leases, when dust was a drug on the market, reflect the situation in the low prices charged for dust to the slag operator.

Let us summarize the slag position at the close of World War II. The slag industry had become a new slag industry—the old tips had largely been worked out. The competition of foreign slag in the

[1] Sir F. Scopes, *The Development of Corby Works*, 1968.

south-east was ended.[1] Both iron and slag technologists knew what was wanted as regards slag quality, and, sometimes, how to secure it. The iron industry was reasonably prosperous, but was using more and more foreign ore of high iron content; it was also introducing new methods of disposing of slag at the furnace.

The post-war period opens with Tarslag out of the slag industry in the west Midlands except at Round Oak; likewise Silvertown in London. Prestwich are anxiously seeking new sources of aggregate to supply their sales connections. Tarmac, the giant, still had great disposal problems on Tees-side and at Corby, and had little interest in securing natural aggregate.

We have seen in the narrative of Tarmac Ltd. that the great anxiety of Dorman Long in the 1954 Teesport negotiations was that the quantity of slag might not be capable of being marketed. Yet in little over ten years from that date, large quantities of old slag were having to be won from the High Tip in order to maintain the output of the plant which Tarmac had built. The first easing of the situation had already started for purely local reasons. Durham County Council for decades had refused to buy aggregate not produced in the county, thereby excluding slag from the south side of the Tees and Northumberland whinstone. Inevitably, high prices and inefficient production resulted. The advent of a new and enlightened County Surveyor brought about a complete change of policy after the failure of persuasive efforts, and opened up Co. Durham as a market for south-side slag.

The main reasons for the turnround were, however, the great general increase in demand and the unanticipated acceleration of the process of using high-quality foreign ore, significantly reducing the make of slag per ton of iron.

Similar events were taking place elsewhere. The chance of obtaining increasing quantities of new slag became very slight. The last opportunity, seized by Tarmac and Prestwich in 1962, was at Park Gate.

Moreover the new disposal methods raised problems of quality of slag, which the reduced quantity per ton of iron did nothing to

[1] J. B. F. Earle, 'English Foreign Trade in Roadstone', *National Westminster Bank Review*, May 1970.

solve. Ladles were dear; slag nearly everywhere was being poured into pits, sometimes into 'furnace pits' without the benefit of any cooling in ladles. Over these problems, the industry has now very largely triumphed. Slag weights per cubic foot have declined, often below the 1942 standard, but it has been proved that the lighter-weight slag is entirely satisfactory and it has for some time now received official customer approval. Yet this lighter slag needs more binder, and when sold per ton, coated slag is more costly than it was. Corresponding advantage, however, comes when it is sold laid per square yard.

As this history closes, with the iron trade still expanding, it seems that slag production is more or less static. All good slag is being used. There are no great reserves—except still at the High Tip at Teesport. The future expansion of the roadstone division of Tarmac Derby must lie in the use of natural aggregate.

This chapter cannot conclude without a discussion of the commercial management of slag disposal by the iron companies. We may dismiss the matter of old slag tips. Their owners' attitude, whether they were iron companies or not, resembled that of quarry landlords. They were normally content to pick a good tenant, fix a minimum rent, and rely on royalties per ton. Freshly produced slag was quite a different matter. In the twenties we find many ironmasters marketing their own slag—notably Stanton, Appleby, Partington (Lancashire Steel), Dorman Long, and Sheepbridge. By the mid-thirties, the atmosphere is very changed. Stanton has always retained the marketing of its blast-furnace slag in its own hands. Appleby took in specialist partners: the others all made leases with slag entrepreneurs, and washed their hands of the problem. It would seem that this flowed from a realization of the complexity of marketing, and from a healthy respect for the sales organization of some of the slag companies, in particular Tarmac and Prestwich, coupled with the long depression in both iron and roadstone. Other iron companies entered into partnership by way of joint companies from the beginning of development. Examples are Guests in South Wales, and Lysaghts at Scunthorpe. The normal form, however, became one of being paid for the slag on the basis of a percentage of the prices being obtained by the slag operators. Negotiations, complicated by the difficulties of cost-accountancy of a by-product, were often long and tedious, but the arrangements worked satisfactorily.[1] The Ford slag at Dagenham was sold by annual tender.

Little reference will be found in the narratives of the slag companies to problems of labour. In fact, the constituent companies of the Roadstone Division have an excellent record as to disputes. For one thing, ever since World War II, the slag industry has been so capital-intensive that its labour force has been comparatively small. It has undoubtedly benefited from its connection with the steel industry with its remarkable tradition of good labour relations. The National Joint Industrial Council for Slag was constituted in 1947, the trades unions concerned being the Iron and Steel Trades Confederation, the National Union of Blastfurnacemen, the Transport and General Workers' Union, and the Union of General and Municipal Workers. The slag employers combined in a Federation, which simplified negotiation. What little difficulty there has been in recent years has mainly arisen from the progressive success of the great generalized unions in recruiting members from the specialized ironworkers' unions. As a result of Transport and General Workers Union influence there has been a lining up of wages and conditions with the quarrying industry.

I have deliberately refrained from incorporating in this chapter matters of a technical character, except those peculiar to slag, like cold asphalt. From the point of view of end-product, it must be emphasized that, as soon as the British Standards Institution in conjunction with the Road Research Laboratory began the issue of standard specifications for road aggregate and coated macadam, granite, limestone and blast furnace slag figured, with minor reservations, as of equal status, though various highway authorities were early to prefer granite and slag for the running surface of their principal roads. We shall therefore avoid redundancy by dealing with end-product specification in the next chapter. Similarly as regards production techniques, the problems of quarry-owners and slag companies are so similar, both as

[1] Scopes, op. cit., pp. 125–6.

regards choice of loose plant, breaking and screening, and the
manufacture of coated macadam that a single description should
adequately cover the matter in a book of this size and nature.

We pass on, therefore, to the subject of quarrying and the
coated macadam industry in relation to the Roadstone Division of
Tarmac Derby.

CHAPTER VI

Quarrying and Coated Macadam

Rough quarries, rocks and hills whose heads touch heaven
It was my hint to speak, such was the process.
Shakespeare, Othello

I could submit the last chapter as a rough sketch for the history of the British slag industry. Tarmac Ltd. alone has played such a predominant part in that industry that the stories of the Group and the trade are inextricably interwoven. It is a very different matter when we turn to quarrying. There was no large quarrying company among the constituents of the Group till 1936. In fact a cynic might well suggest that a better chapter-heading would have been another quotation from *Othello*—'a chronicle of small beer'. We can allow ourselves a smile at Peter Preston the First's £3 a week, the struggles of J. N. Cuthbert in Ayrshire, and those of the first merger in the Rowley Hills.

It would be quite wrong to suppose that our grandfathers did not know how to quarry on a large scale. There were many substantial quarries in Edwardian times; the fact, about which I shall hope to say more in Chapter X, that limestone and chalk are chemical raw materials, brought about quarrying on a big scale for ironworks, cement manufacturers, and others with a substantial tied market. The railway companies, too, were quarrying on a large scale for track ballast. Even some of the famous roadstone quarries were already sizeable, notably Clee Hill and Enderby inland, and Penlee and Penmaenmawr on the coast. None of these are in the

Group; I cannot emphasize too much how late was the serious impingement on the trade of its constituents.

The leading textbook of those times is *Practical Stone Quarrying*, Greenwell and Elsden, Crosby Lockwood, 1913. In over 500 pages it describes techniques of development, extraction, blasting, cableways, and the like, which soon dispel any notion of our predecessors' ignorance. There is a great deal of reference, it is true, to continental and American practice, but many excellent installations in Britain are described. However, there is no reference to the Group quarries except for Caldon Low—then a railway company quarry—and, surprisingly, to Middleton-in-Teesdale, where the Ord and Maddison paddle-mixer for tarmacadam has engaged the authors' attention.

As the demand for chippings developed, so did the answers of mechanical engineering. E. W. Blake had patented the jaw crusher in the United States in 1858 and by 1862 Marsden was selling it in this country. Other famous engineering firms produced improvements and modifications, notably Broadbent, Baxter, and Goodwin Barsby. Normally cubing rolls were used in conjunction with the Blake-type jaw crusher, but by 1911 Hadfields were manufacturing the Symons vertical disc crusher, and 1913 saw the hammer or impact crusher in great demand from the Lightning Crusher Company. After this, in the twenties, came the new epoch in secondary crushers with the Nordberg and Telsmith machines, and still later the development of the impact breaker from a secondary or tertiary machine into a primary machine, restricted to aggregates of low abrasion.[1]

I have spent a little time on this account of quarry engineering in order to demonstrate that it was not practical difficulties which inhibited our predecessors' expansion. It was not really lack of demand, either. In 1922 Tarmac made over £84,000 profit and Tarslag over £30,000. Compare these slag figures with the best of the quarrying companies, Ord & Maddison and Cliffe Hill, both around £15,000 average in the immediate post-World War I period. The explanation lies largely in the geography of

[1] *1862–1962: One hundred Years of Stonebreaking*, 'Macadamer', unpublished MS.

CECIL MARTIN IN RETIREMENT

JOHN HADFIELD

quarrying, and in the limited outlook of the quarry-owners.

Before proceeding to discuss this, mention must be made of the serious effect on the industry of today of the very narrow outlook of these Victorian and Edwardian quarry-owners. With apparently inexhaustible reserves of stone, having regard to their small annual outputs, they opened their little faces at the points most immediately convenient. No geological research was done, and very little thought given to planning development. The result, seventy years later, of the errors of siting of plant flowing from this has frequently been to put the modern roadstone quarrying industry to great capital expense which could have easily been avoided.

None of the present-day quarry enterprises of the Group except Inverkeithing lie on the seaboard. Penlee and Penmaenmawr, with their wide option of markets by sea, were large enterprises when the Tarmac Derby constituents were tiny. Inland quarries are always faced with competition within a mile or so; they are therefore limited in the size of the market that they can command though the situation is ameliorated if the quarry is on the edge of the quarrying area, when all the local quarries, as in Leicestershire, may have a large market in non-stone bearing country to share between them. Of the two successful inland roadstone quarries already mentioned, Enderby was in this happy position, but the exceptional standing of Enderby and Clee Hill must, I think, be attributed to very good sales management of the type exhibited by D. G. Comyn of Tarmac.

Such was the economic situation of the roadstone quarries. What of the quarry-owners? The narratives tell of either small family companies or of small syndicates of local people, as at Rowley Regis and New Northern Quarries. Generally their horizon did not stretch beyond their own locality, and if it did, they did not know where to find extra capital. They were at a great disadvantage compared with a concern like Tarmac, with leading industrialists in the chair, and City connections.

The roadstone quarry-owners of the twenties, when the demand really started in earnest, were, moreover, in technical trouble. The early files of the *Quarry Manager's Journal*, first published in 1918, are full of the difficulty of making tarmacadam, especially with granite, as reliable as or equal in quality to tarred slag, and the discussion continues for years. The orders were there—in May 1920 a great shortage of tarmacadam is reported—and the atmosphere was most optimistic,[1] but the solutions of proper grading and controlled binders took long to be found. It would seem that the quarry-owners did little to help themselves, at any rate in Scotland, where at a meeting of the Institute of Quarrying as late as January 1929 it was stated that there was not a quarry making tarmacadam in that country which had yet got a drier.[2]

So we arrive at the end of the twenties. A mass of scattered units, in fierce competition with one another, with little recourse to new capital, and with the depression descending upon all British industry. New ideas about the roadstone quarrying industry were, however, developing.

In 1929 the British Quarrying Co. Ltd. was formed, with Sir Henry Maybury as chief executive. This was a combination of successful quarries in various parts of the country—Kent, Shropshire, south Gloucester, etc., and did not really deal with the problem of local competition, though it provided means of raising fresh capital, and the beginnings of modern industrial organization in the scattered trade. At the same time, the prophets were preaching rationalization and price co-operation. Mr. L. J. Bancroft from Yorkshire, and Mr. Hugh McCreath, Senior, in Glasgow were outstanding in trying to break down the individualism of the past. So, of course, were Cecil Martin and Major in the slag industry. Gradually, the quarry-owners began to seek salvation through co-operation. As a matter of history, though, the author must recall the animosity, suspicion, jealousy, and sheer bad manners which so often characterized the early meetings of the trade.

By 1934 the associations were widespread and working reasonably well to improve prices. In some places, as in London, some rationalization was taking place. It required, however, John Hadfield's conception and execution of the Derbyshire Stone merger to give a further lead to turning roadstone quarrying into

[1] 'The Coming Boom', *Q.M.J.*, Feb. 1923.
[2] *Q.M.J.*, Jan. 1929.

'big business', and, in the end, to lead to the Tarmac Derby Roadstone Division of today.

World War II did a lot of good to the roadstone industry, both in the quarrying and slag sections. Demand for airfields and other military works was high. The habit of co-operation grew apace under the compulsion of patriotic fervour. The Ministry of Transport largely left the industry to organize supplies itself, involving dozens of concerns, and it responded well. The introduction of the mechanical finisher showed the leaders, at least, what they had got to do about their production arrangements, and a rush of new construction started as soon as the war was over.

The history of the fifties as regards the Tarmac Derby quarrying interests is an interesting economic study. The price associations were working so well that the compelling urge to local mergers, as with Derbyshire Stone in 1935 and Roads Reconstruction in Somerset in 1934, was quite absent. A heavy increase in demand, as Chapter I has shown, hung fire till the end of the decade, and the successful companies reacted in different ways. Derbyshire Stone, had, through Hadfield's modern management methods, done very well, but, broadly speaking, was still a local concern in Derbyshire. Vertical integration did begin to appear, as the personal connection of Hadfield with Ragusa led to the absorption of Amasco, but the general policy of the Board was one of complete diversification, starting with the acquisition of the chemical firm of Baird & Tatlock in 1959. The Hughes and Cuthbert families took an entirely different line, that of horizontal expansion, in the first case into Lancashire and North Wales, in the case of Kings throughout Scotland. Ord & Maddison, New Northern Quarries and Cliffe Hill continued their conservative policies, though Peter Preston the Third was beginning expansion into ready-mix concrete and gravel. A new, and finally overwhelming, factor was beginning to appear. The limitations of slag supply had led to the purchase of Rowley Regis by Tarslag. By the end of the decade Tarmac, too, had begun to realize their own aggregate problem.

Ever since the war, two legislative trends were exercising more and more influence—land planning and estate duties. Planning was making it very difficult to open new quarries, especially in granite, and was consequently enhancing the value of the old ones, provided they had large reserves. In Leicestershire, for instance, despite desperate efforts, no planning permissions for quarrying had been granted since the early fifties. Limestone was admittedly less of a closed shop, especially in the north of England, but reserves of suitable stone held on freehold or long leasehold terms were not easy to find. The slag industry tried to exploit the position with the help of the Iron and Steel Board. An acrimonious meeting took place at the Ministry of Transport in 1947, at which Ministry of Supply officials were present, at which the argument was advanced that it was contrary to the national interest to open fresh quarries while unsightly tips of good slag remained; it had little effect on the Ministry of Transport.

So far as estate duties were concerned, private family companies were liable to disintegration on the passing of the principal shareholders, and a further factor in the desire to realize their assets in the form of cash or marketable securities was that the age of the industry had now reached a stage where control was frequently in the hands of the third or even the fourth generation.

So we enter the sixties, the story of which I have set out rather bleakly in the Tarmac Ltd. narrative. It records one absorption after another; some of them at apparently very high prices, often over ten years' purchase of profits. The principal objective, however, was always the reserves of aggregate, often to maintain or expand existing markets, and quite different financial criteria were applied from those governing ordinary capital expenditure on plant and vehicles. Exceptionally, in the story of the sixties, is the acquisition by Derbyshire Stone of Kings, an imaginative piece of 'old-fashioned' horizontal expansion.

We are now recording the making of a modern large enterprise. The problems of the sixties, and doubtless, of the early seventies, were those of welding a heterogenous collection of quarrying and slag enterprises—for we can no longer deal with them separately—into a homogeneous entity. The story vibrates with administrative and personnel difficulties—and opportunities. The management of Tarmac Ltd. was by the wise decisions of the late fifties, mainly in the hands of a young generation, for flexibility and ability to learn modern industrial methodology rapidly were absolute necessities. The annual capital budget had, of course, been long established, but its subservience to a five-year plan and to calculations of discounted cash flow, and even more sophisticated techniques, were novelties which required much education and even 'brain-washing'. Sophisticated cost accountancy became the order of the day. There were, inevitably, plenty of personal problems, in particular with those companies with a residue of family management with a distinguished sales record. The widespread geographical distribution of the production units and markets enforced decentralization, while the fact that, like peace, the British roadstone business is essentially indivisible, pulled in the opposite direction. Under the circumstances, a growth of overheads tended to absorb the obvious economies of rationalization, and top management had to weigh up the cost of modernization such as a Group planning department and a computer and of central roadstone services, such as engineering design and quarrying expertise, against their predicted advantages.

It is probably fortunate that nearly three years elapsed between the last major acquisition of the old Tarmac Ltd.—Hillhead Hughes—and the 1968 merger. By the time fresh problems of rationalization had to be faced, a great deal had been accomplished by J. M. Beckett and his colleagues. Moreover into the merger, George Henderson brought much experience of the same sort, not least in the successful rationalization of Kings and Amasco in Scotland since 1965.

Though negotiation has not been so smooth within the National Joint Industrial Council for the Roadstone Quarrying Industry, formed in 1919, as in the case of the corresponding body for slag, this has largely flowed from organizational complications. There are seventeen districts within the N.J.I.C. for roadstone, and meetings have often involved nearly thirty trade unionists (though all from the Transport and General Workers' Union and the Union of General and Municipal Workers) and a large number of quarry-owners. Minor details of district problems were inclined to inhibit the smooth negotiation of national base rates, yet it must be said that the industry has a good record of solving its labour problems without undue trouble. Capital intensification has, just as in the case of slag, played its part. For example, Derbyshire Stone reported that in the ten years 1948–58, the value of plant per quarry-worker increased from £648 to £2,278. When one considers the laborious nature of hand quarrying, this sort of development obviously appealed to worker and employer alike.

The employment of aggregate in roadwork takes many forms. Large quantities of quarry waste are required for 'fill'. The sub-base is a great consumer, and nowadays the specifications for the stone or slag used are quite severe. Chippings for surface dressing have been a continuous and important market from the earliest days of tar spraying. Granite and gravel have been the principal selection in this field, but slag and to some extent hard limestone have played an important part. One of the difficulties which the constituent companies, and of course the industry as a whole, has had to face has been the fluctuating change of fashion as to size of chippings, as a result of theory and practice in dealing with the problems of skidding. Fillers, sand and coarse aggregate are fundamental ingredients in road asphalt.

Large quantities of slag and limestone (and occasionally granite) have been used in mechanically-laid road bases in the form of wet-mix. The story of its introduction into motorway construction after satisfactory use on an airfield is well set out by Mr. James Drake, the County Surveyor of Lancashire, in a lecture delivered in 1957.[1]

As an aggregate for concrete road construction, whether in surface or foundation, the Group's products have been constantly in demand.

However, the spectacular use of roadstone, slag, granite, and limestone, is in the manufacture of coated macadam, and it is to this activity of the Group I shall devote the rest of this chapter.

Coated Macadam

The first mechanical mixer seems to have taken the form of an Archimedean screw which drew the dry or dried macadam up

[1] F.C.M.I. Lectures, 1957.

through tar and discharged the finished material at the top. As we have already seen, it was ten years before Tarmac, the protagonists, contemplated using anything but hot new slag for their material, and so drying difficulties did not exist for them. In the quarries, however, the primitive drier was a perforated plate heated by a fire beneath. The next two developments foreshadow the modern coated macadam plant. Ord & Maddison, whose engineering manager was Mr. W. A. Hiscox, afterwards himself to become a leading mixer manufacturer, were certainly making and installing a paddle-mixer by 1907. The drum drier was an early improvement on the hot plate, though the author saw the latter still in commercial use in the early thirties at Forest Rock in Leicestershire.

By the time World War I was over, however, the sudden increase in demand for tarmacadam was engaging the attention of the designers of the engineering industry, especially those interested in contractors' plant. A popular plant was made by Ransome Machinery Co. in sizes up to 30 tons per hour output, and consisted of a batch-type combined drum drier and paddle-mixer. This type of plant was offered in small sizes, even as little as 4 tons per hour, and obtained a ready sale from highway authorities, to the detriment of the nascent coated macadam industry. One cogent reason for 'do it yourself' coated macadam was the relief of chronic local unemployment. A real economic factor, soon to disappear as delivery by lorry took the place of delivery by rail, was the relatively high freight rate charged by the railways on the finished material. Most of these small local authority units had disappeared by the end of the thirties, economies of scale having triumphed over the inefficient production of these very labour-intensive machines.

Broadly speaking, the changes in plant design between the wars were fundamentally improvements on the Ransome type of machine. Inclination of the mixer tips to cause the material to travel through the mixer, and double paddles were soon introduced, involving much more power and a strengthened axle.

Better, and above all, quicker mixing was the result, with beneficial effects on cost. One of the great bugbears of the early manufacturers was overheating of the aggregate—probably at the back of many of the *cris-de-cœur* in the *Quarry Manager's Journal* of the early twenties to which I have already referred. Arrangements therefore were introduced to give time for cooling and for the aggregate to 'sweat' before the binder was introduced. While the engineering firms were quite capable of introducing means of weighing grades of aggregate and measuring accurately quantities of binder into coated macadam plants, such refinements were most unusual in this period, though the same engineers had perforce to incorporate them in their asphalt plants. Tarmacadam specifications were of the loosest description—a typical county council order for the late twenties would read '700 tons bottoming and 70 tons topping'. As long as the bottoming's top size did not exceed $2\frac{1}{4}''$ and the topping $\frac{3}{4}''$, and the material consolidated under the roller, the material was acceptable. Naturally brand sales flourished, and here Tarmac had a great advantage over its competitors. What the total market was in 1924 is impossible to ascertain, but Tarmac's figure of over a million tons must surely represent more than half of it. Slag also, for many years, as we have seen, maintained a genuine technical advantage over the granite quarries, many of whom, even if they had learnt how to introduce properly dried aggregate at a correct temperature into the mixer, had problems of grading which the use of slag solved with little trouble. These happy-go-lucky days were, however, coming to an end and the issue in 1938 of B.S. 802 started a new era of proper scientific control of manufacture, involving, for those who took it seriously, a great deal of capital expenditure. I cannot leave the old days, though, without paying a tribute to that great craftsman, the mixer-man, who by experience and diligence produced large quantities of coated macadam of first-class quality by eye. It is sad, in a way, to have to record that, in the automatic plants by which within ten years the old machines were being replaced, it was found better to employ operatives who had never mixed before, and were prepared to believe in the validity of their dial readings.

With World War II, the great days of the coated macadam industry were to begin. Petroleum residual bitumen had by then been extensively used for over ten years, and despite its price disadvantage against tar, was popular for three main reasons. First, it was much more satisfactory in coating 'difficult' granites. Secondly, its resistance to oxidation gave the surface more life without dressing. Thirdly, together with slag, it had solved the cold asphalt problem. Bitumen was helping coated macadam to look at hot rolled asphalt in the face. Next, the exigencies of war brought mechanical spreading to this country years earlier than would probably have been the case. In the same context, the airfield demand conditioned manufacturers and road engineers alike to the idea of large daily outputs. The industry itself was nervous and apprehensive of both concrete and asphalt competition, and it was the apparent leaning of the Ministry of Works to concrete roads for housing that triggered off the formation of the Federation of Coated Macadam Industries in 1944. It was very soon, however, that the roadstone industry realized that the money was in coated macadam, made properly in modern plants. The asphalt industry was full of difficulties, already detailed in Chapter IV, and which will be discussed further in Chapter VII. There were frequent shortages of cement. Coated macadam provided a cheap and durable road crust, and that was what the customer wanted. The very success of the fifties was, maybe, dangerous to the Group's constituents. Tarmac Ltd. themselves and Hillhead Hughes paid very little attention to dry aggregate on the one hand, and to asphalt on the other. Nevertheless 1944–64 was a remarkable period for coated macadam from large static plants, and technically it is closely associated with automatic mixing, and in particular with the designs of Stothert & Pitt Ltd. These plants had the great merit of being able to produce material to the now proliferating range of end-product specifications which were being called for without any significant waste of mixing time. Generally ten sizes of aggregate were available to the operator.

The first Tarmac Ltd. rebuildings after the war did not incorporate automation, or even satisfactory weighing and measuring devices. Tarmac Ltd.'s production management was conservative and it is noteworthy that they lagged behind their competitors in plant design until R. G. Martin obtained a position of authority. The new or reconstructed plants at Ettingshall,

Acklam, Corby, and Scunthorpe were distinctly old-fashioned, and compared unfavourably with the Stothert and Pitt type installations at Whitwick, Vaynor and Horrocksford. With the rebuilding of Irlam and Skinningrove under R. G. Martin's direction, the whole attitude changed. The Tarmac mixer—one of the company's own design—which was significantly slow in operation, now was everywhere replaced. The almost atavistic objection to driers— shades of Purnell Hooley!—gradually disappeared, and the introduction of the principle of surge piles contributed to continuity and lower costs. There were men of the same engineering outlook as R. G. Martin in many of the constituents, notably Hobday of Derbyshire Stone and Witchell of Crow Catchpole.

At first, it appeared that coated macadam had received a setback when, after trying out a cold asphalt surface on the first motorway, the Preston By-Pass, the Ministry of Transport determined on a hot rolled asphalt policy for the surfacing of major construction. In 1960, however, a specification for dense bitumen macadam and dense tarmacadam for use in roadbases was published. Recognition was quickly accorded to the outstanding advantages to be derived from the use of this material and the Ministry of Transport specified its use as an alternative to hot rolled asphalt for the full depth roadbase construction wherever there was risk of subsidence, as in Durham and Midland coal-mining areas. The materials were also frequently used elsewhere. The merits of dense coated macadam moreover brought it into the base course of many major constructions, though hot rolled asphalt was invariably specified for the wearing course of such roads. The unsuitability of lean concrete base as originally used on the M1 soon became obvious and led to the introduction in 1962 of what has become known as the composite road base comprising 7″ of lean concrete of greater strength topped with 3″ of dense bituminous material. All these factors provided substantial orders for coated macadam. The total demand on F.C.M.I. members therefore increased rapidly after 1959. The relevant figures are set out below; the recent plateau in cold asphalt is noticeable, but the increasing demand for dense base course and dense roadbase is almost phenomenal.

F.C.M.I. and A.C.M.A. Statistics
(millions of tons)

Year	Coated macadam Total	Cold asphalt	Dense base course and dense road base
1945	2·64		
1950	4·87		
1955	6·18	not available	
1960	8·08		
1961	8·81		
1962	9·78		
1963	10·10	1·02	0·89
1964	12·45	1·22	1·51
1965	12·92	1·25	2·12
1966	13·22	1·19	2·41
1967	15·87	1·37	3·63
1968	17·20	1·38	4·81

Coated macadam and hot rolled asphalt are now welded together in one trade organization, and nearly every sizable concern manufacturing the one material is also making the other. As our story closes, there is a greater growth in asphalt than in coated macadam, and the competition of concrete is strengthening. I must comment further in the next chapter, but with roughly a third of the coated macadam trade in their hands, members of the Tarmac Derby Roadstone Division may look back at Mr. James Ward's tarmacadam plant in north Lancashire in 1898, and take pride in the achievement in this department in no more than a man's normal lifetime.

CHAPTER VII

Road Asphalt and Surfacing

Fac tibi arcam de lignis lavigatis:
mansi unculas arca facies, et bitumine
linies intrinsicus et extrinsicus

Vulgate, Gen. 6:14

To understand the part which road asphalt has played in the
history of the Tarmac Derby Roadstone Division it is very
necessary to bear in mind five fundamental differences from coated
macadam.

The first is that asphalt, whether natural or derived from refinery
bitumen has a multiplicity of uses in protection against water.
Our chapter-heading describes an early specification in marine
architecture. The Bitumen Products Division's history is not within my terms of reference, but much can only be understood by the
realization that asphalt, for the two main purposes of roadwork
and building as well as the minor ones, has throughout a great deal
of the story been made in the same factory.

The second point is that road asphalt is regarded by highway
engineers as normally superior to coated macadam for the surfacing
of heavily-trafficked roads. Experience of twenty years without
maintenance is not uncommon, and highway engineers in certain
circumstances are quite prepared to pay considerably higher
prices for asphalt than for coated macadam.

The third fundamental has largely contributed to the second.
The technology applied to asphalt has been far greater and is of

far longer history than that applied to coated macadam. For example the first British Standard Specifications for two-course and single course asphalt (hot process) were published in 1928, ten years before B.S.S. 802 for tarmacadam. Moreover these specifications were based on the pioneer work of Clifford Richardson in the United States and afterwards in this country in the nineties of last century. His work was of so remarkable a character that the mixture which he recommended after his visit to England in 1896 is very similar in all respects to the sand carpet specified in the current B.S.S. 594. His precepts were followed rigidly by the leading asphalt companies, and they backed their products with massive laboratory control decades before coated macadam manufacturers adopted scientific methods. They generally offered five years free maintenance, and made great efforts to produce a voidless material. The result was that right up to the end of World War II, highway engineers regarded coated macadam not only as unsuitable surfacing material for the heaviest traffic, but as a second-class material generally. It was one of the objects of the formation of the Federation of Coated Macadam Industries to change that image, but it took twenty years to do it. The merger of the trade associations in the Asphalt and Coated Macadam Association in 1967 may be regarded as a recognition that both materials are first-class products, some specifications being more suitable for certain road work than others.

The fourth great difference is that little road asphalt is sold by the ton. The concomitant of sound technology in manufacture was expertise in laying, and this was particularly important prior to the days of mechanical spreading. Therefore the asphalt industry developed a 'contractor' psychology in contra-distinction to the 'manufacturer' outlook of coated macadam. I submit that this has contributed to the commission of certain errors of commercial judgement in the asphalt industry. Outside the cities there has been a proliferation of small plants, because every order is a 'job', and generally a 'job' requires a plant to itself. Again the civil engineering contractor is by nature a migrant, concerned with local successes, which in the asphalt industry has too often diverted attention from the need for permanently established sources of aggregate. There are, of course, two sides to this, and the absence of the 'contractor' mentality may well partly explain the slowness of Tarmac Roadstone Ltd. to establish themselves firmly in asphalt in the sixties.

Finally, and from the point of view of an economic historian, most important of all, is the fact that road asphalt is significantly more expensive than coated macadam. This is inevitable for a number of reasons. First is the cost of binder per ton of material. B.S.S. 594, is, in the eyes of foreign engineers, an extravagant specification, but British highway engineers are most reluctant to agree, because of their experience of the long life of the material. A prophet is not without honour except in his own country, and the Americans' practice has deviated far from their own Clifford Richardson's precepts. In fact, in talking to Americans about asphalt surfacing, it has to be remembered that they are nowadays referring to nothing very different from bitumen macadam. Secondly, though the difference as compared with coated macadam is less marked today, the cost of technological control has to be paid for and it is very high. Thirdly, because of the lack of miscellaneous sales on which the coated macadam industry thrives, asphalt plant is far less utilized than coated macadam plant. The 'job' conception prevails, and it has been customary to estimate annual output of an asphalt plant by the application of a multiplier of 1,000 to the hourly rated figure, instead of 2,000 as in coated macadam. Both labour and depreciation costs are seriously affected by this. Finally, attention must be drawn to the much higher cost of heating in an asphalt plant.

One exception to the general proposition that road asphalt is more expensive than coated macadam must be noted. It is technically much more satisfactory to utilize asphalt methods of manufacture than those of coated macadam when gravel is to be used as the coarse aggregate. At the same time gravel is not very acceptable for heavily-trafficked roads. The ideal circumstances are consequently in East Anglia, where gravel is in good supply while other natural aggregates and slag are particularly expensive through transport costs and where traffic conditions are not exacting. Although plenty of gravel asphalt is used elsewhere, its hold on counties like Norfolk and Suffolk must be fully appreciated.

Though we are not strictly concerned with the early days of road asphalt in this country, except that the Neuchatel Asphalte Co. Ltd. supplied natural asphalte to the Val de Travers Company, development will be clearer if I give some brief account of them.

The original city street asphalt, as supplied in 1869, consisted of $\frac{1}{2}''$ mastic on which was superimposed $2''$ of compressed rock asphalte. This was followed, as an alternative, by the mastic form of pavement, using the same rock asphalte but with additional bitumen, forming a plastic mass laid by floats. The considerable advances of the next forty years largely flowed from American research. Finding European rock asphalte far too expensive, the Americans moved on to what they called 'sheet asphalt', consisting essentially of stone, sand, dust or filler and bitumen (so confusedly termed 'asphaltic cement') which originally was largely Trinidad or other lake asphalte. As the years went on, the petroleum companies began to market the residual bitumen from their refining process, and both mastic asphalt and hot rolled asphalt, to use the English term for 'sheet asphalt', were manufactured with refinery bitumen or a mixture of it with Trinidad lake bitumen. In the case of mastic, so largely used for building purposes, this involved the use of specially selected limestone powder.

So in 1913, a year mentioned in two of our narratives, we find Chittenden & Simmons using Trinidad asphalte in hot rolled work in Kent, and Highways Construction being formed to market the residual bitumen of the Mexican Eagle Oil Company.

It is frequently not appreciated how early the petroleum companies entered the road bitumen market. This is partly due to the great boost that was given to the activity by the appointment of Mr. J. S. Killick, one of the country's leading road engineers, as Bitumen Manager by Shell-Mex and B.P. in 1922. Killick was a sales manager of great ability, and with the massive organization behind him, sales of bitumen for the manufacture of asphalt, coated macadam, cold emulsion, and other road materials thereafter increased very rapidly. Esso Petroleum and Berry Wiggins and Co. (generally supplied by Gulf Oil) joined in and for forty

years the three concerns maintained a virtual monopoly of refinery bitumen sales in the United Kingdom, though there were a
small number of independent concerns such as our own William
Briggs. Sales of refinery bitumen exceeded those of road tar, and
displaced Trinidad Lake Asphalte to a considerable extent, though
this remarkable material maintained a great deal of its former
hold, despite the expense connected with the inability to handle it
in bulk. It is also to be noted that the price of bitumen was well
maintained, always higher than that of tar, and generally much
higher than on the Continent.

By 1924, the Neuchatel Asphalte Company was established in
the British market, and all the constituents of Amasco were at
work, though it must be remembered that Ragusa concentrated
on the building market. The importation of European rock
asphalte became more and more uneconomic, and Trinidad Lake
Asphalte became the normal source of the natural material, while
refinery bitumen was increasingly utilized. Highways Construction made an effort to utilize rock from Avignon from 1936, with
their speciality of 'rock-non-skid', but the costs defeated them
and their imitators in the end. While the old companies held on to
the limited mastic trade, new competitors arose in the main field
of hot rolled asphalt, slowly dispelling the mystique with which the
old companies had successfully reinforced their salesmanship.
Our narratives record Tarslag and Kings in particular entering the
market, which always at a premium over coated macadam, except
perhaps in East Anglis, was fiercely competitive though we note
an improvement towards the end of the thirties. We can note
diversification into coated macadam; one of the objects was better
employment of the laying gangs, and Neuchatel became substantial tar-paviors. The narratives indicate a half-hearted aggregate acquisition policy, except in the case of Neuchatel, where the
pursuit of aggregate continued after World War II, particularly in
Scotland.

The narratives of Neuchatel and of Amasco itself give a picture
of the fifties which I need not elaborate except to emphasize the
sudden dearth of asphalt orders from the American Air Force
and from the 'Attlee rearmament policy' which occurred in 1955,

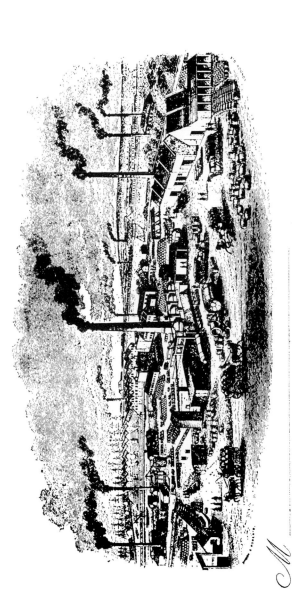

ℳ⸺

Bought of Edwᵈ Catchpole & Sons

Tar Distillers & Manufacturers

of Naval Varnishes, Coal Tar, Naphtha, Grease & Asphaltum.

EARLY DAYS OF TAR—*ca.* 1865

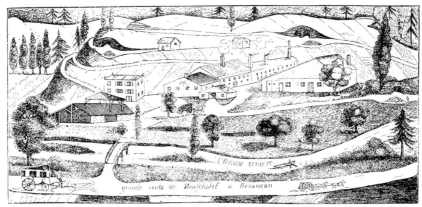

Domaine, Exploitation et Usine des Mines d'asphalte de *la Presta près Travers* (Suisse).

Domaine, Exploitation et Usine des Mines d'asphalte de *Chavaroche près Annecy* (Savoie).

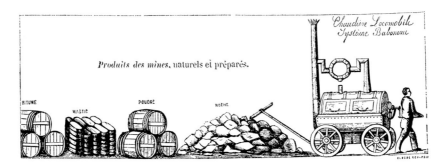

EARLY DAYS OF ASPHALTE:
THE VAL DE TRAVERS MINES AND FACTORY

leaving a market hiatus till the construction of M1 in 1959 opened up a new large and growing source of demand. By 1962 Amasco was a successful, large and well-organized company and demand for road asphalt was growing at a greater rate than that for coated macadam, a fact which had brought further formidable competitors into the field, such as Tarmac Ltd., A.R.C., and Roads Reconstruction. Road asphalt was dominating the demand for the surfacing of motorways and other new construction, and problems of organization were much to the fore. A divisional organization, operating from headquarters based on a large city static plant, did not fill the bill, when very large if spasmodic orders were being obtained for motorways, and a separate set-up became necessary. The undoubted fact that, in the last six successful years of our Amasco story, motorway work did not contribute its proper share of profit partly flows from these administrative difficulties, and partly in its earlier stages from exaggerated ideas of the prestige to be obtained from connection with the great new development.

Surfacing

As described above, the asphalt industry has always included the laying of the material as part of its normal functions. This has not been the case with coated macadam, where practice has been diverse. Tarmac originally laid a certain amount of its product, as illustrated by the picture facing page 47, but when a civil engineering department was established in 1929, the surfacing activities were transferred to it, and since the new department was operated almost as a separate business, for thirty years it was regarded in effect as another customer for production. It was not till 1960 that, surfacing being transferred to Tarmac Roadstone Ltd., the laying of coated macadam began to be integrated with the production. Of the companies primarily concerned with slag, Tarslag and Prestwich always laid a considerable part of their production, but Crow Catchpole avoided the complication, while Silvertown used a sister company. With the exception of Kings, who always seem to have had a 'contractor' mentality, the quarrying companies also refrained from laying activities. We have seen already how valuable to

KRM

Tarmac in the laying both of asphalt and coated macadam were their Tarslag and Prestwich acquisitions, and how they were helped in the rapid development of surfacing as an integral part of their business by the experience of these two concerns.

We must now enquire into this diversity of policy. We have already seen that in the case of asphalt, there were very real technical reasons for combining manufacture and laying. It is certainly the case that highway engineers, who could not be expected to be responsible for the operation of the highly skilled technique involved, fought shy of laying asphalt themselves. Coated macadam was quite another matter; its laying by hand is not a difficult art, and widespread gangs contributed to the size and importance of the highways engineer's department within his local authority. As the demand for coated macadam rose, however, the authorities were forced to let the contractor in, because of establishment limitations. There were in London and some of the large cities, a number of small tar-paving concerns, based primarily on private sector work or on public work outside the scope of highways, such as for education and military authorities, and they were ready to expand. At the same time, as we have seen, the asphalt companies were often eager to fill in their programme gaps with the laying of coated macadam. Civil engineering contractors —and, by the forties Tarslag and of course the Tarmac civil engineering division regarded themselves in that category—also were glad to increase their turnover by this relatively easy means.

It can be seen, therefore, that as the constituent companies became more and more coated macadam manufacturers, they contemplated a great number of contractor customers, as well as their original clientele of highway authority customers. Most of our constituents therefore took the view that they did not wish to alienate their contractor customers by competing with them in laying, which in any case did not seem so profitable as manufacturing.

The alternative view, adopted by the minority, was that surfacing provided a tied market for manufacture. All sorts of compromises were utilized to try to get the best of both worlds. Derbyshire Stone was on very friendly terms with its contractor shareholders.

Silvertown utilized a sister company. Tarmac and Crow Catchpole in partnership in London tacitly split the business, Tarmac concentrating on the highway authorities and Crow Catchpole on the contractors.

In the fifties a new factor began to enter into policy. It had been relatively easy to organize the price-fixing associations as far as supplies were concerned, but the minor details of surfacing contracts made any arrangements in regard to them far more loose and easier to evade. To take on laying jobs was a short-sighted but immediately profitable means of expanding production without debasing supply prices, and the temptation was too strong for some producers to resist, though it was self-defeating in the long run.

It also became obvious to producers, who were expending very large sums of capital on increasing and improving their production, that minor sums spent on laying equipment, which in any case could largely be hired, were likely to provide a satisfactory return.

We, therefore, arrive at the present situation, which seems both logical and likely to be successful. The Division has a large surfacing organization, expert in asphalt as well as coated macadam, which carries on a business intended to yield a good return on capital by itself, while providing an assurance of a substantial, though only partial, market to the static plants. The organization has sorted itself out into a combination of divisional and national control, which seems to afford the best administration in the circumstances. I hope, however, that the account of earlier years will indicate that present policy flows from much very hardly-won experience in a field full of pitfalls.

CHAPTER VIII

Rail, Road, and Sea

Dirty British coaster with a salt-caked smoke stack
Butting through the Channel in the mad March days

Masefield, 'Cargoes'

The fundamental realities of geology and of population distribution, on which I have touched in the Introduction, determine
the overwhelming importance of transport costs to the roadstone
industry. All the narratives of Chapters II and III, and to some
extent of Chapter IV, must be read in the knowledge that management was for ever trying to reduce its transport costs, at any rate
relative to competitors. Of the delivered price even of coated
macadam, with the large cost of binder involved, one-quarter to
one-half is represented by the cost of various phases of transport.[1]
The proportion is normally therefore higher with uncoated material.

The great roadstone market of south-east England, where the
factor is nearer one-half than one-quarter, is also very accessible
by sea, particularly via the Thames Estuary, so again the narratives
continually reflect the struggle between internal transport and
shipping, not only coastal but from the Continent as well. The
flexibility of sea-transport has brought about an almost kaleido-
scopic picture of the Division's sea-borne trade, with the rare
phenomenon of imports being replaced by exports. Its complexity
deserves a section to itself, so readers will find that this chapter
is divided into two parts, the first being comment on the rail and

[1] F.C.M.I. Lectures, 1957. J. B. F. Earle and following discussion.

road transport used by the constituent companies, and the second a description of their sea-borne trade.

Prior to World War I, distribution of the Group's roadstone over any appreciable distance was almost entirely by rail. Perhaps the only substantial exception was at Rowley Regis, where the proximity of the excellent canal system of the west Midlands, constructed in 1770–1830,[1] brought about a large amount of barge traffic. Short hauls, and the clearance of rail wagons to road site, were very much a matter of horse-traction, though we observe the light lorry beginning to be used, and Cliffe Hill actually buying lorries as early as 1914. World War I gave a great boost to the development of road motor transport, and by the time it was over, the horse and cart was fast giving way to the motor vehicle for delivery over short distances. There was a flood of new entrants into the road haulage business, many of them ex-service men buying ex-service lorries from the Ministry of Munitions with the help of their war gratuities.[2]

All through the twenties, however, long-distance transport of roadstone was by rail, and the narratives continually refer to the buying of main line wagons by the companies which served the markets of East Anglia, London, and the South. The normal practice of county councils in these areas was to buy at railhead. Their tender forms set out every goods station in the county, to each of which a price was required. Either with the county council's own transport or by hiring, the trucks were unloaded by hand and the material conveyed to the road jobs. However, with binders by no means perfectly adapted to winter weather, and long delays on rail, it can be imagined that disasters were not infrequent on, for example, January deliveries from the Midlands to the east coast of Norfolk.

This method of delivery applied often to quite short distances. For example, the works of Crow Catchpole at Newhaven never had road access, and from 1925 until World War II, everything was delivered by rail to the stations of East Sussex.

As large users of the railways, the roadstone industry and

[1] C. I. Savage, *An Economic History of Transport*, Hutchinson, 1959, pp. 17 ff.   [2] Savage, op. cit., p. 133.

Tarmac in particular, were deeply concerned with railway rates. The Railways Act of 1921, which merged the large number of railway companies into four main groups, after the idea of nationalization had been rejected, substituted a new system of charging for the one that had been in force since 1891–92.[1] A new classification and schedules of standard rates were projected, but negotiation was so complicated and lengthy that the new classification was not published till June 1923, and the schedules were not completed till 1927. The narratives (see, for example, New Northern Quarries) indicate combination in the roadstone industry so as to participate effectively in the negotiations. D. G. Comyn of Tarmac played a leading role. He skilfully got uncoated slag into Class I by arguing that if it were not sold it would have to be dumped. He thereby gained a slight advantage over his quarry-owner competitors, uncoated stone being in Class II.

By the end of the twenties, the transport atmosphere was changing rapidly. The switch to road deliveries can be traced in one narrative after another. Tarmac, who had amazed themselves by keeping going with road transport at Ettingshall in the railway strike of 1924, started buying lorries for delivery of finished material in 1929. New Northern Quarries were forced into large capital expenditure at Silverdale in order to have adequate road access. Derbyshire Stone, who inherited large numbers of railway wagons on its formation in 1935, soon sold them, and applied the proceeds to buy road vehicles.

The change-over to road vehicles was greatly assisted by improvement in design, especially in tipping gear. Though a number of self-discharging and tipping wagons are described at quite an early date,[2] effective gear seems to have been introduced into the roadstone trade from 1926 onwards. Originally of the pinion and quadrant type, for some years vertical screw tipping gear was used, first hand operated by two men, then by a power take-off attached to the gearbox. The year of break-through was, however, 1936, with the introduction of the swash-plate hydraulic pump,[3] which

[1] Savage, op. cit., pp. 102 et seq.
[2] H. J. Butler, *Motor Bodywork*, W. R. Howell & Co., 1924, pp. 296 ff.
[3] *Hydraulics at Work*, Golden Jubilee Issue, Edbro Ltd., 1969.

could be used on vehicles having a payload of 8 tons. The industry was demanding larger and larger payloads and with tipping gear developing to match the increasing size of lorries, the revolution was completed.

By the time World War II was over, the delivery of coated macadam by rail was largely a thing of the past. Railway embargoes during the conflict had accelerated the process. We see Cliffe Hill's light railway from the quarry to Bardon Hill sidings abandoned to become perhaps a relic for future archaeologists.

The railway era of coated macadam was not without its merits. Continuous manufacture over double shifts was perfectly possible, and formed an agreeable contrast to the restricted manufacturing hours of today. Customers were content with cold material, were accustomed to giving long notice of their requirements and to waiting patiently for material to arrive. One great disadvantage to the Group companies was the necessity of financing the purchase of railway wagons for the delivery of coated macadam, for, after the passing of the Railways Act of 1921, the railway companies refused to carry the material in common-user wagons.

So far as inland transport is concerned from the end of World War II, the management of the constituents was applying itself to the best utilization of road vehicles. What proportion to buy, and what to hire; how to manage transport, whether as a separate organization or as a function of production; what type of vehicles to buy; whether they should be powered by petrol or diesel engines, and so on. There was a considerable amount of disagreement on some of these matters.[1] Tarmac organized their road transport under management quite detached from that of production, so it was practically a separate business. The advantages claimed were that this produced the most efficient running of the owned vehicles, and that the yardstick of return on capital invested could be easily ascertained. Others, Crow Catchpole being particularly vocal, dissented. They argued that the maximum profitability of the owned vehicles by no means meant that the profitability of the business as a whole was maximized. The semi-independent transport manager tried to obtain a maximum number of loads

[1] F.C.M.I. Lectures, 1957.

daily for the owned vehicles. It was argued that service suffered, production was lost, and the hired vehicles, not being treated fairly in the queue for loads, cost more than they would when they were loaded in the order of their appearance. Critics pointed out that the internal transport accounts were based on the so-called market rate, i.e. that charged by outside hauliers, and that this rate could be appreciably lower if there were no discrimination. The Thames Tarmacadam Association of Tarmac and Crow Catchpole operated a transport undertaking, but the partners' views were so divergent that it was wound up, but not before it had been demonstrated that hired haulage rates at Hayes, controlled by Crow Catchpole, were 10 per cent lower than those at Irlam, a plant of similar size operating under similar conurbation conditions.

As the fifties advanced, diesel power replaced petrol, but only slowly. The capital cost was in total formidable. The semi-independent transport manager was tempted to buy a larger number of petrol vehicles for the same money as a smaller fleet of diesels, another example of total profitability not necessarily being enhanced by increased profitability of the transport department.

Not the least of the blessings of the sophisticated cost-accounting developed in the sixties has been the ability it has given to top management to assess the real performance of the large amount of capital invested in road vehicles.

The railways, gradually excluded from the direct coated macadam traffic, came back into that trade in the mid-thirties in a different way and have played a prominent part ever since. The growing size and complexity of the London market made it impossible to give proper service without local manufacturing units, and the new transport problem was how to supply these local units with aggregate. The railways soon demonstrated that by conveying uncoated slag or stone, particularly in train-loads, they could enable the London plants to deliver coated macadam over Greater London at prices highly competitive with direct coated macadam supplies by road from the Midlands or Somerset. The traffic was ideally suited to the railway set-up, and engaged the interest of its top management. In contra-distinction to road and sea haulage, railway rates on bulk aggregate fell during the period 1955–68. Traditional good relationships with the railways helped our constituents, notably Tarmac, Derbyshire Stone, Cliffe Hill, and Crow Catchpole. The major competitor, A.R.C., was slow to follow suit, though in its acquisition of Roads Reconstruction, it found a company very much alive to the possibilities of rail transport in its coated macadam organization not only into London but into Sussex also.

The Sea-borne Trade in Roadstone

The supply of roadstone to south-east England by sea, both from the Continent and from sea-board quarries such as Penlee, Penmaenmawr, and in the Channel Islands, had been a regular feature of the industry since before the turn of the century.[1] None of the Division's constituents, however, played any part in it until after World War I. During the war, as we have seen in the Tarmac narrative, the Acklam works supplied a considerable proportion of the 300,000 tons delivered to Dunkirk from the Tees and used by the British Forces in France.[2] Though D. G. Comyn was attracted to the sea-borne trade in 1918, the Tarmac Board were averse and the real pioneers so far as the Division is concerned, were Crow Catchpole, closely followed by Silvertown Tarmacadam and Tarslag, though it is to be noted that Inverkeithing, the Division's only British sea-board quarry, was being developed by Tilbury Contracting for supplies to East Anglia and elsewhere.

Crows' first venture was the shipping of slag into the southeast from the Ruhr, mainly from the Thyssen works and from Duisburg. They set up a depot at Vlaardingen, near Rotterdam, where the Rhine-lighters of slag were discharged and reloaded into ships for England. Plants were set up in quick succession at Rochester, at Newhaven and Littlehampton, in London, at Dover, and Ipswich. Crows also shipped slag in the mid-twenties from

---

[1] J. B. F. Earle, 'English Foreign Trade in Roadstone', *National Westminster Bank Review*, May 1970.
[2] 'English Foreign Trade in Roadstone', op. cit.

Port Clarence and other points on the Tees. Silvertown followed suit in 1926 by importing slag from Belgium via Ghent, and Tarslag operated a works at Dagenham, supplied both from Stockton and Belgium. In 1926 Crows also turned to Belgium, and started taking considerable quantities of slag from the Charleroi and Liège areas. Wage rates in Belgium were low, and the nationalized Belgian railways offered phenomenally cut rates to the ports, so that Germany began to disappear from the picture. By 1929, the situation started to change. Dorman Long and Bolckow Vaughan merged, and began serious shipping of slag from the Tees. By this time, too, the political campaign carried on by the Institute of Quarrying, with the help of Sir Cooper Rawson, M.P., against the import of foreign aggregate was becoming effective and Crows and Silvertown began buying heavily from the Tees. Moreover the financial crisis ending in the devaluation of sterling in 1931 dealt the Belgian producers a fatal blow.

Yet another foreign source of slag was, however, to appear. The Société Metallurgique de Normandie built large slag plants at Caen in immediate proximity to the Ouistreham Canal, and so, being right on the sea-board, triumphed over the exchange problem, and shipped considerable quantities to England, of which Crows took a large part, especially to Newhaven and Littlehampton.

As the Tarmac narrative describes, by 1933–4, Tarmac had acquired the agency for all the blast furnace slag shipped by Dorman Long, had themselves large shipments from Acklam, and were established on waterside premises at Poole, Crown Wharf, and Shoreham. They were gradually establishing working arrangements with Crows and Silvertown, and by the end of 1934 were supplying a number of other London manufacturers with sea-borne slag.

The cost of freight had deterred the Tarmac Board from following up D. G. Comyn's shipping proposals of 1918, but by this time, sea transport between the Elbe–Brest limits had become very cheap—3s. per ton freight being a normal figure from the Tees, east Scotland, France, and Ireland to south-east England.[1]

[1] 'English Foreign Trade in Roadstone', op. cit.

Coated macadam based on sea-borne aggregate was dominating the market in Sussex, much of Kent, and much of London, Essex, and East Anglia. Apart from Caen and Crows' widespread purchases of sea-borne igneous rock, which are described in some detail in their narrative, Tarmac had become the major supplier of sea-borne aggregate for the coated macadam trade, and were themselves considerable manufacturers, though Crows were outstanding. Cecil Martin and Earle soon found common ground, and were leading the organization and to some extent rationalization of the coated macadam trade in the south-east.

I must not leave this description of the Division's sea-borne trade between the wars without referring to the quantities of coated macadam, mixed at or near the port of shipment, which were shipped. The Tarmac narrative describes the lack of success in this direction in 1928, but Crows persisted for some time with German slag mixed in Holland, and even marketed material from Dalmellington in Ayrshire, shipped through Irvine. The delay to ships by weather or by causes flowing out of their previous voyages proved in the end to be insurmountable.

World War II, of course, brought all commercial shipments of aggregate to an end for five years, but, as soon as it was over, the manufacturers concerned started to pick up the threads again, as detailed in the narratives. Tarmac and Crows were working together very closely, and before long had concentrated their Sussex manufacture at Shoreham. Rochester was abandoned, what market that remained being supplied from Deptford. Crows' sea-borne slag supplies came almost wholly from Tarmac, though a little was still taken from Caen. The new factor in the sea-borne trade of the Division to the south-east was the development of sea-borne limestone from Devonshire, under the leadership of Mr. W. R. Northcott, and we see both Crows and Silvertown developing a coated limestone trade of some magnitude.

The conditions governing the sea-borne trade were, however, fundamentally changed, and it soon became apparent that England would no longer be an importer of slag, but a substantial exporter, and in this rather extraordinary development Tarmac played a leading part.

I do not want to duplicate what I have already published on this subject,[1] and for full details including geological comment would refer my readers to the monograph noted. A brief outline and explanation is, however, necessary for the completion of this chapter.

The relative costs of sea-transport as opposed to road and rail, had moved against shipping, which, though technically far removed from Masefield's coasters of our chapter-heading, had in particular to face phenomenal increases in labour costs. This trend, especially compared with railway rates, has brought about a shrinkage in the last twenty years of our story of the economic area of the south-east favourable even to British sea-borne slag. Mendip limestone has advanced its frontiers. Rail-borne slag and stone have increased their proportion of the London market. Though brought about by other circumstances, the disposal of Deptford has done the Division no harm. It is the really remote works, Ipswich and Dover, which have seen continuous prosperity in these circumstances.

This factor of loss of competitive power by sea transport would in any case have inhibited foreign competition after the war, but there were other more direct factors at work.

Demand for road materials, especially in the actual theatre of war, as around Caen, was growing more rapidly in the continental countries of north-west Europe than in this country for fifteen years after the war ended. The home market therefore absorbed the aggregate, and in north Germany the domestic slag supply became inadequate by the late fifties. Other markets at one time relying on German supplies, such as Denmark and Holland, faced a hiatus. A variety of uses for slag other than as aggregate was developed, particularly in Germany and Belgium, more rapidly than in this country. The stage was set for exports of slag—and of igneous rock, though apart from Inverkeithing the Division had no means of supply of the latter.

The situation therefore was one of economic limitations of coastal shipments on our side of the North Sea, and a growing demand on the other side. The Tees-side slag trade, by the early

[1] 'English Foreign Trade in Roadstone', op. cit.

sixties entirely in Tarmac's hands, was looking for orders, and could rely on great reserves, especially in the 'high tip' at Teesport.

The regular flow of exports started in 1950 with shipments of slag dust to Denmark for that country's developing demand for cold asphalt, but it was, and is, north Germany where big quantities could be sent. Tarmac showed considerable enterprise in the matter, setting up their own sales office (now at Hamburg) and selling direct without the interposition of agents, who have always been employed by their slag and granite competitors. The best year to date was 1964, and when this history closes, decline had taken place due to a variety of causes, mainly effective Scandinavian and Irish competition. Scunthorpe also came into the trade, despite the disadvantage of having to haul to the port of Flixborough on the Trent, and obtained valuable orders for granulated slag from cement manufacturers at Lübeck. Orders in Holland were, however, very difficult to obtain; the Ruhr is too near (by Rhine-lighter) and the great Belgian quarries of Quenast and Lessines were effective competitors.

When this history closes, the situation as to slag on the Continent was similar to that in this country, as described at the close of Chapter V, save that it seems that the 'high tip' at Teesport is unique in north-west Europe.

Since this chapter is intended to throw light on matters of transport, I would add to this very brief account of the export trade that the fact that there is a substantial import of potash from north Germany to the Tees has often made it possible to negotiate freight rates from the Tees to north German ports more favourable than those to the south of England, where return cargoes are hard to find.

CHAPTER IX

Co-operation in the Roadstone Industry

Not forsaking the assembling of ourselves together

Heb. 10:25

As we have seen, the roadstone industry began to take shape in the last thirty years of the reign of Queen Victoria and in that of her successor, King Edward VII, and it consisted of a large number of small local businesses. My personal recollection of its managers over forty years ago is that they were intensely individualistic. I will attempt an explanation of this attitude. These people, and their immediate forebears, had, like myself, been brought up in the psychological atmosphere of *laissez-faire*, preached so effectively by Jeremy Bentham around the middle of the nineteenth century, and in the biological revolution of Darwinism, popularly epitomized in the expression 'survival of the fittest', which was without justification extended to apply to nearly every human activity. Moreover, the political doctrine of free trade, highly suitable to an island where industrial development was well in advance of its competitors, still dominated fiscal policy when world conditions had fundamentally altered. The two great emotions which unite men, religion and patriotism, were of course present, but the first swayed industrialists between the two poles of Christian compassion, typified by Lord Shaftesbury and the Factory Acts, and the more unpleasant traditions of Puritanism, while patriotism, perhaps as a consequence of Anglo-Saxon racial lethargy, only became effective in times of grave peril, when it amazed the world.

I must now trace the change from the attitude of mind where the competitor was a deadly enemy to that which largely dominated the thinking of the managements of the roadstone industry by the end of the thirties.

To begin with, the broad picture of the past which I have painted requires a good deal of examination and criticism in detail.

My researches lead me to believe that the earliest association of a formal nature in the industry was the Leicestershire Granite Association, of which Cliffe Hill was one of the original members. It was in existence in the early years of the twentieth century, and seems to have originated in the need for common action in matters concerning the safety and welfare of quarrymen, and of action required to conform to the various Mines and Quarries Acts and Regulations. It was not however till 1930 that any reference occurs in the well-preserved Directors' Minute-Books of Cliffe Hill to co-operation on prices, though one suspects that the occasions of formal meetings for other purposes may have been utilized for informal discussions thereon. One may speculate on the reasons for the early date of some co-operation in Leicestershire; it seems that the quarry-owners in that country were largely in the landed gentry class, as compared with elsewhere, and may consequently have found the concept of mutual benefit more easy to accept than others. They also found a far-seeing leader in Mr. John German, a highly-respected estate agent.

Similar, but more shadowy bodies, seem to have existed prior to the end of World War I, for there does not seem to have been grave difficulty in the formation in 1918 of the National Federation of Granite and Roadstone Quarry Owners, and of the Central Association of the Lime and Limestone Industry at about the same time. These two bodies formed a co-ordinating body, the Federated Quarry Owners of Great Britain, in 1919, and the industry largely under the inspiration of Mr. S. MacPherson, set up an Institute of Quarrying, which published monthly *The Quarry Manager's Journal*, the files of which have provided me with much information about the period between the wars.

We see the limestone body in action in 1921–3 from the New Northern Quarries narrative. This was in presenting a common front in the negotiations on rate classification under the Railways Act of 1921, of which I have written in Chapter VIII. The most sustained effort for mutual assistance in these early years would seem to be that of the Institute of Quarrying against the importation of foreign roadstone. The campaign certainly was in full force from 1926–33. An influential Member of Parliament interested himself in this, and by public and private representations, Sir Cooper Rawson, M.P. helped the Institute to persuade many highway authorities in south-east England, to insert 'British origin' clauses in their roadstone tenders.[1]

The Asphalt Roads Association dates from the same era, so we see three of the facets of the Tarmac Derby Roadstone Division, granite, limestone, and road asphalt, organized nationally for some purposes of mutual benefit at a fairly early date. Formally, and it would seem also even informally, price arrangements did not enter into the discussions which the meetings of these bodies facilitated. There cannot be certainty on this point, but it must be remembered that from 1919 to 1924, nearly every one of the narratives of the constituents reveals prosperity and expectation of boom conditions. An atmosphere of prolonged depression and crisis, and even of disaster, was necessary before the old attitude of mind which I have described could possibly be abandoned in favour of comprehensive co-operation.

It was from the slag industry that the first initiatives came. The Tarslag narrative tells of Mr. Major's efforts, and of the rebuffs of D. G. Comyn of Tarmac, who, one may speculate, was deceived by his own success in the field of sales management. As we have seen, Comyn's retirement through ill-health coincided with the collapse of Tarmac's profits in 1926. The appointment of Cecil Martin as Managing Director of Tarmac, while prices, through unbridled competition in a contracting market, were continuing to fall, led to conversations with Prestwich, Tarslag, and others. A new prophet appeared, Mr. L. J. Bancroft of Brookes Ltd., who had marked success in convincing slag entrepreneurs in the Midlands and Yorkshire of the need and practicability of some arrangement on prices.

[1] J. B. F. Earle, 'English Foreign Trade in Roadstone', op. cit.

NEWCASTLE

1887.

LONDON

1851.

Ord & Maddison, Ltd.
QUARRY OWNERS,
& AGRICULTURAL ENGINEERS.
TELEGRAMS:—"ORDE, DARLINGTON."

Darlington.

190

We beg to acknowledge
with thanks, the receipt of your favor
enclosing   for £

LASTLY.

No intoxicating drinks shall be brought upon the Quarries without the sanction of the Inspector, and no workman shall be admitted to the works in a state of intoxication. The workmen are strictly cautioned against the use of profane and improper language, and the Owners earnestly request them carefully to abstain from a practice so degrading to themselves and offensive to others. Any workman using foul and improper language, subjects himself to dismissal from the works.

◄ 1874

RELICS OF
ORD &
MADDISON

1869
▼

CONDITIONS

On which every

employed at these Q

Every man entering th
Maddison at these Quarr
a copy of the Rules, Regu
Hiring, by which he will be
be returned on his leaving the

ORD & MADDISON, AGRICULTURAL
ENGINEERS,
DARLINGTON, AND WEST BRIDGE DEPOTS,
MIDDLESBROUGH,
Supply on the shortest notice Best Household and other
Coal and Coke ;
Carboniferous and Magnesian Lime and Limestone ;
Whinstone and other Road Material ;
Flagstones, Slabs, Curbstones, Building Stones, Gravel,
Sanitary Tubes, Chimney Tops, Drain Pipes, Bricks, Tiles,
Flooring Squares ;
Guanos and Chemical Manures ;
Also, Metal Castings and Agricultural Machinery of every
description.   22

J. LAW. GREENWOOD STREET,
CORPORATION STREET, MANCHESTER,
will forward a RECIPE for making GOLD PLATING
LIQUID. Warranted to stand the the test of any chemical
reparation, for dipping Watchguards, Rings, Lockets,
rooches, &o.—Terms, 13 stamps.
Extract from the *Chemical News*, February 23, 1862

SHE
at th
the Black

MUS
BO
good paper
3d. and 6d
CONCERT
words), in
Books, at
Hotel), 10

D A

THE BOARD OF SHEEPBRIDGE MACADAMS LTD., DECEMBER 1957

R. G. Martin, C. C. Wilson, R. W. Prestwich

H. Fletcher, L. J. Hodgkiss, C. Martin, J. Hadfield, J. Prestwich

As the years of depression continued, and conditions worsened, the movement for co-operation on prices gathered strength. It was not without its sympathisers in the ranks of the county surveyors. Generally compelled by standing orders to buy at the lowest price tendered, they experienced the normal troubles to be expected in supplies from near-bankrupt concerns. Quality and service both deteriorated. County Councils which themselves owned quarries were well aware from their own accounts that current prices were inadequate. Though the formation of price associations was kept discreetly in the background, the fact that they existed was known to the county surveyors, if not to members of the County Councils, and in general not resented unless they suspected exploitation, which, in fact, was rarely even possible because of the presence of a considerable amount of non-association competition.

Before I give some description of the price associations to which the constituent companies of the Division were parties, I must pause to recall the conditions under which roadstone suppliers had to tender for the annual requirements of highway authorities.

Over a period of several decades, forms of tender remained fundamentally the same, and so did the standing orders flowing from them. Devised originally to gain the most advantageous prices for small quantities of goods by persons of Victorian up-bringing who distrusted their own underpaid officials, they had to be adhered to rigidly under penalty of rejection.

With rare exceptions, the tenderer was required to bind himself to supply whatever quantities might be called for, without any information (except from his experience) as to the probable demand. The tenderer was not informed at what seasons of the year the demand would be heavy or light, and in any case the weather risks were his. The only thing suppliers did know from experience, was that provided it was not snowing heavily, a greatly enhanced rate of demand would arise in March—'mad March' in Masefield's poem, previously quoted—because the authorities had to get rid of their annual supplies of money by the thirty-first day of that month. Incidentally, even county surveyors often did not know how much money was available till a late date in the financial year. County treasurers would keep money 'up their

LRM

sleeve', the Treasury would release money unexpectedly to the Ministry of Transport, who offered grants to County Councils who could both match them from the rates (justifying the treasurer's prudence) and use them before the end of March. Moreover, the cost of late winter snow clearance, included in the same vote, was unpredictable. Next, the tenderer had often to submit one price for a very large area involving significant differences in transport from the point of production. He would make a profit if the roadworks happened to be near his quarry; he might well make a serious loss if it was thirty miles further away.

Perhaps, however, the most damaging factor in the system, was the disinclination of authorities to reveal the results of tenders until long after they were submitted. Tenders for supplies for the year starting on 1 April would be called for from the previous mid-November to mid-February. Tenderers might have to commit a significant proportion of their annual production to a large County Council in a tender sunmitted in mid-November, and be unable to learn what the result was till the end of January, by which time they would have submitted nearly all the other tenders.

Finally, the doctrine of the 'lowest price' prevailed.

When price associations started in 1928–36, these conditions were all present, but were not nearly so vital to the producers as they were in later years, because the business was so much smaller. The original urge was simply to get prices generally on to a profitable basis. It was as quantities increased, and particularly as mechanical spreading greatly enlarged the daily requirements of jobs that association procedures were modified and shaped to cope with the rigidity of an archaic tendering system.

The first formal price associations date from 1928. These were slag bodies on the north-east coast, and in London. 1929 saw the formation of the North Wales Tarred Macadam Association covering slag, granite, and limestone. In 1930 slag producers in the Midlands and Lancashire respectively got together, and before long extended their numbers to include natural aggregates. As we have seen in the narrative of the formation of Derbyshire Stone, an association in Derbyshire was running, unsatisfactorily, in the early thirties. The two great West Country associations in the

Mendips and south Gloucestershire date from 1935–6 and the same date saw the start of the Central Scotland and East of Scotland bodies.

The association with the largest number of members throughout its history was the South Wales and Monmouthshire Quarry-owners' Federation. Started in 1933, a considerable part in its formation was played by the trades unions, who were concerned that, as a result of the tendering policy described above and the consequent great irregularity in the distribution of orders, unemployment and excessive overtime working were existing side by side. As a result, it seems that service was exceptionally bad in the South Wales area.

Geographical gaps in the association system were rapidly filled; important bodies were the Scunthorpe and Derbyshire Group and South Midland Association, both confined to slag and both formed in 1934.

As the whole of this chapter has sought to demonstrate, the idea of price co-operation in the roadstone industry involved a complete change of mind in its managers, and this did not come about either easily or quickly. Association agreements provided easy escapes, and some of the earlier ones soon broke up, only to reform on experiencing a repetition of the old disasters. On reforming, they improved; people were learning. 1937–40 was a period when the original experiments were over, the parties had begun to know and even in some instances to like one another, and it will be useful to have some account of the variations in constitution which had evolved from local conditions and difficulties.

One fundamental difference between association price structures is contained in the words 'equal prices' or 'cover prices'. Of course every member wanted what he thought was his fair share of the business. The cover price, or allocation, principle was used as a means of deliberately arriving at this result. It had the further merit, from the producers' angle, of disguising the presence of the association. It prevailed generally throughout the south of the country from the Mendips to Essex, where belief in the merit of sales connections was weaker than in the Midlands and north, where there was much equal price tendering, which worked pretty well when it was found that the large County Councils would generally split their orders when confronted by equal prices from two or more suppliers. These immediately pre-war associations were also marked by fairly complicated money transfers. These flowed out of quota systems with compensation, as in the Mendips —later abandoned—or as a result of a profit-sharing system based on theoretical bare costs. The latter was introduced in the London Slag Federation primarily to solve the problem of persuading Highways Construction (q.v. narrative, Chapter IV) to abandon shipping slag to London. It worked well, and had imitators.

The South Wales body was, from the first, distinct from the other associations in acting to some extent as sales agent for its members. This developed further later as we shall see.

The new spirit of co-operation was spreading through British industry generally. All had experienced the depression; all were determined that these agonising years for industrialists should not return. In the roadstone industry the spirit manifested itself in further ways other than mere price association.

The joint venture, or partnership, began to flourish in the thirties, and Tarmac were great protagonists of the idea. The ventures which were started flowed from differing immediate causes, as set out in the narratives, which contain numerous examples.

1936 saw the formation of two important bodies, the British Limestone (Roadstone) Federation, and the British Slag Federation. They had nothing to do with prices; their objects were propaganda, technical improvement, and joint action in dealings with Government departments and bodies of the same nature. Together with the old-established British Granite and Whinstone Federation, formed under a slightly different title in 1918 (see above), they constituted a means of national representation of the roadstone industry. The interesting thing is to note *when* they were formed, i.e. after some years of co-operation on prices, and after the re-education that that experience had brought. A feeling of unity in an industry was now triumphing over the old individualism.

So we arrive at World War II, and in acute national emergency, the new spirit of co-operation was both enhanced and utilized fully by the Government. In December 1939 the War Office turned to the national associations in their need to recruit specialists for the Royal Engineers in large numbers in the shortest possible time, and their success in fulfilling the requirement was acknowledged by the formal thanks of the Army Council. It was, however, after the fall of France, and the imperative need to construct airfields in Britain on a very great scale and in a very short time that the benefits of experience in mutual working showed themselves fully. The Ministry of Transport soon found that the industry could organize supplies on a voluntary basis far more efficiently than could be done from Whitehall. A Stone Control was set up by statutory order, but the civil servant administering it worked hand-in-glove with the national associations, a Joint Committee of the three Roadstone Associations being formally constituted to advise. By the time the war was over the new spirit of the thirties was regarded by the managers of the roadstone industry as the normal relationship between their still competing companies. So it was that in 1944 three men, Cecil Martin of Tarmac, W. R. C. Hockin of Roads Reconstruction, and Stewart Mitchell of General Asphalte Co., the surfacing contractors, brought practically the whole of the coated macadam industry, both suppliers and layers, into a new body for technical advance and propaganda, the Federation of Coated Macadam Industries. The author was the first Chairman, and was followed by a distinguished line of successors chosen from the chief executives of leading companies. The F.C.M.I. increasingly prospered under the direction of two eminent road engineers, Mr. A. J. Lyddon and Mr. W. Mervyn Law, until its merger with the Asphalt Roads Association in 1967 to form the Asphalt and Coated Macadam Association. It may be here noted that the three aggregate federations likewise have now also merged in the British Quarrying and Slag Federation. Mergers are the keynote of the sixties, as associations were of the previous decade. This is, however, to anticipate.

Our narratives will have shown that the earlier mergers of companies now in the Division generally flowed directly from hard times and cut-throat competition. This was the case in 1900 at Rowley Regis, in 1935 in Derbyshire, and in 1957 in the formation of Amasco, and applied, outside our immediate story, to the Roads

Reconstruction merger in Somerset in 1934. The benefits of large-scale operation and of rationalization were clearly present in the minds of those combining in Derbyshire Stone, but were not their primary motive in embarking on permanent combination. It is to be noted that Derbyshire Stone's early acquisitions were all three in the direction of eliminating local competition.

The first mergers which seem to have been actuated by motives corresponding to those of Tarmac's great expansion of the sixties seem to have been Hillhead Hughes in 1935, an ambitious exercise in growth, and Tarslag's acquisition of Rowley Regis, motivated largely by hunger for aggregate.

When the war was over, and after an interval in some cases for picking up the threads, there ensued a period when the price associations in the roadstone industry worked with maximum smoothness. There were of course numerous local difficulties and plenty of outside competition to restrain exploitation of the now established spirit of combination, but nevertheless one tendering season followed another with some confidence that when all the bargaining was over, a member would be assured of a reasonable tonnage at a reasonable price. Elaborate money adjustments were eliminated as unnecessary in a give-and-take atmosphere. The agency idea flourished in South Wales, being particularly effective in Monmouthshire.

Into this comfortable state of affairs, the Conservative Government's Restrictive Trade Practices Act of 1956 fell like a stone into a pond. The twelve years that follow up to the end of this history, which coincides with the passing of the Labour Government's Restrictive Trade Practices Act of 1968, are characterized by two things. The first was the determination of the industry that, within the new law, the benefits that it had enjoyed from association working should be preserved as far and as long as was possible. The second was the later realization that any measures to this end could only be temporary and that the respite gained must be employed by organizing the industry on a satisfactory but fundamentally different basis.

The industry's leader in combining to take what measures were legally practicable was John Hadfield, supported very effectively by Cecil Martin. A Committee of Roadstone Producers was constituted by the price associations and had high hopes of passing the roadstone bodies unscathed through the Restrictive Practices Court. First-class advice was obtained from economists as well as from lawyers and accountants. Three associations were selected for reference, Leicestershire Granite, Central Scotland Whinstone, and the Cold Asphalt Group of the London area. Reference to the pleadings of the Respondents in these cases show the conviction of the industry that the removal of the restrictions would result in less capacity and consequently inability to meet surge demands, less capital re-equipment, and a decline in service. It was argued too that in the long run, owing to the elimination of the weaker concerns, higher prices would result.

However, while industry generally optimistically believed that the intention of Parliament had been only to eliminate exploitation of industrial co-operation, it soon became obvious that the Judges construing the Act were of the opinion that very few price associations would pass through the necessary gateways. On Counsel's advice, therefore the three Associations concerned gave undertakings to abandon their specified practices and 1961 marked a big change in attitude.

I have emphasized above in this chapter the damaging disinclination of some highway authorities to publish speedily the results of annual tenders. Lawyers advised the industry that, though they could not continue their price-fixing agreements, they could resort to price-reporting agreements. A chain of centres with this in view were thereupon set up, generally covering the areas of several of the old price-fixing associations. Normally the reporting of prices to one another through a Centre was done prior to their submission, but following the rather devastating judgement in the so-called 'Galvanized Tank' case, it was advised that this procedure was too near to the edge of peril, and it was abandoned in favour of reporting prices after submission, which was still helpful in an orderly disposal of works capacities. The 1968 Act and the Orders made under it restrict the interchange of prices by agreement unless they have been previously published to customers.

In the twelve years since the passing of the 1956 Act, however, the structure of the industry had been radically altered. I will conclude this chapter with some general comment on the merger and acquisition wave of 1959–68. Chapter I has shown that the total expenditure on roads increased threefold in those ten years. Ambitious and enterprising public companies naturally turned towards obtaining a large share of what was manifestly a growth industry. Those involved can be divided into three main groups. The first is the 'internal' group—those already largely occupied in the industry. Apart from Tarmac, whose activities have already been detailed, and the Derbyshire Stone–Amasco–Kings combination, the best example of this group is A.R.C.–Thomas Roberts– Roads Reconstruction, although they ended up in the second group—that of other participants in the extractive industries, to whom the technical lay-out was familiar, English China Clays, Redland Holdings, Hoveringham Gravel and more recently Consolidated Gold Fields in their acquisition of the greatly expanded A.R.C., being the leading exponents of this group's policy. The third group, that of industrial holding companies, is best typified by Thomas Tilling Ltd., who have acquired significantly large quarrying interests in their variegated portfolio.

The result, and the process is obviously not yet complete, is that power, and consequently commercial policy, in the roadstone industry is concentrated in the hands of a few very large public companies. For example, the nine medium or small concerns occupied in the Leicestershire granite industry in 1950 are today represented by four industrial giants, and one private company. I need not multiply detail; the same development is to be seen in important areas of aggregate production as diverse geographically as the Mendips, the Scottish industrial belt and Co. Durham and Tees-side.

The giants respect one another's abilities—and with modern management, are looking continually for a respectable return on capital. They are not likely therefore, arbitrarily or short-sightedly to cut prices. It may well be that the industry's price structure is more soundly based than it was in the comfortable days of the early fifties.

CHAPTER X

Non-roadmaking Activities

The lime manufactured from this stone is valuable as an agricultural manure. It is used as a flux in smelting iron-ore, also for purifying gas for tanning, and for other chemical purposes.

From the Ord & Maddison entry in the catalogue of the Great Exhibition of 1851

As, in order to preserve the balance of this book, I have tried to deal with non-roadmaking activities in one chapter, readers will find that it has been necessary to combine comment and description with the Group's history in the various markets dealt with under the sub-headings.

1. Chemical and other special uses of Limestone, Burnt Lime and Chalk

The extract from the Great Exhibition catalogue that forms the chapter-heading would seem to indicate that the principal markets contemplated in 1851 by Ord & Maddison fell within this sub-heading, and it is, of course, the case that the use of limestone or burnt lime in the reduction of iron-ore and as a correction to the acidity of the soil dates back to time immemorial.

The narratives of Chapter III make many references to the supply of limestone and burnt lime to the *iron and steel industry* and they still form an important part of the sales of the Derbyshire limestone quarries, in particular Topley Pike and Middle Peak. The story of the acquisition of the latter quarry from Stewarts &

Lloyds Ltd., shows how the output is built round this market. Here, however, I must note how the qualities of limestone differ within quite a small geographical area. The siliceous cap of Middle Peak and nowadays Intake stone generally are not acceptable as a flux in iron-making, and have been diverted into the aggregate market.

When it comes to the *agricultural ground limestone* market, the quality of the stone is of less importance. We see Derbyshire limestone from various quarries, Cheddar 'black-rock', Foss Cross oolite, Aycliffe dolomite (primarily quarried for chemical purposes), Louth chalk and Minera limestone all being utilized. This freedom of choice to the consumer, together with the 'stop–go' policies of the Ministry of Agriculture as to subsidy, and the presence of large numbers of merchants, has contributed to the somewhat disappointing nature of the results achieved after the high hopes of 1938.

Where we see the advantage of a very pure limestone, in the case of Middleton over 99 per cent $CaCO_3$ and a phenomenally low iron content, is in supplies to the *glass industry* in the manufacture of white flint glass. Other quarries can supply a flux for the manufacture of amber glass, but though the history of the supply of limestone from the Group's quarries to the glass industry certainly dates back to 1920, today the only quarry in the Group which can meet the most exacting specifications is Middleton.

The big development in the cultivation of *sugar beet* came after World War I, and supplies from the Group's quarries of limestone as a remover of impurities dates back to the same period. This is an important and regular trade, suffering from the difficulty that the material is only required from end-September to mid-January. The sugar refineries are mainly situated in the East Anglia area, which has always given Derbyshire quarries a transport advantage over their competitors; purity is again an important factor, and Middleton stone is particularly in demand.

I might detail further uses of limestone as a chemical. An American publication[1] gives an exhaustive list, to most of which

[1] *Industrial & Agricultural Uses of Limestone*, J. J. Berliner & Co., New York City.

the Group's quarries have contributed, but I will pass on to its use in *fillers and powders*.

This market has been, and is increasingly, of considerable importance. With three asphalt companies among its original constituents, Derbyshire Stone Ltd. naturally had a very big trade in mastic powder. The first mills for this purpose were set up by Constable Hart & Co. and Greatorex about 1928, and Matlock can still be regarded as the home of the mastic trade.

The supply of ground limestone powders for other purposes by the Group's quarries in Derbyshire really dates from World War II, and the shortages of chalk whiting caused thereby. Though on the reappearance of adequate supplies of whiting, the new market contracted severely, well-directed sales efforts restored and expanded it, so that the Group's limestone powders now sell in all manner of markets, rubber, flooring compounds, carpet backing, and ceramics being a few examples. Moreover there has been some export. Fineness of grinding, purity of stone and whiteness are all important factors in success in this market.

In the *building trade*, apart from its contribution to the general supply of aggregate which I discuss below, another 'from time immemorial' activity in which the Group has participated is in the supply of lime for sand lime or calcium silicate bricks. Contributions have been made to cement manufacture, to special white aggregate for white concrete, and to roofing chips and the like.

Many other uses have been found for limestone fillers and powders, some, such as colliery dust and cattle food, have been abandoned for economic reasons. Although the Group's quarrying of limestone for all these purposes now approaches 2m. tons per annum, it must be appreciated that even when to the Group figures are added the similar productions of its competitors the total is far less than the consumption of the country in these directions. The iron and steel industry, the cement industry, and the heavy chemical industry have all acquired very large quarries in order to supply their raw material requirements, and I have already noted in Chapter VI that together with the railway companies quarrying for track ballast the early development of large-scale quarrying was brought about by these interests with their substantial tied markets.

2. Aggregate Generally

Building and civil engineering are, of course, enormous consumers of aggregate. The preparation of sites for both industrial and house building operations may involve great quantities of 'fill' and all of the Group's types of stone and slag have a continuous history of being used for this purpose. A spectacular contract, such as the Shell Petroleum job of 1964 at Teesport recorded in the Tarmac narrative, catches the eye, but throughout the century of this study, demand for this purpose has gone on. In the nature of things, this type of business is not of high profitability, but it has often been of great value in quarry development with indirect benefits difficult to quantify. Moreover the clearance of old slag tips may set free land for development; this was a most important factor in obtaining planning permission at Jarrow.

It is, however, as an aggregate for concrete, that the processed production of the Group's constituents has found its largest market apart from roads.

I have talked about the twentieth century as the 'automobile age'. From another angle, it might be called the 'concrete age'. As engineers learnt and applied the principles of reinforced concrete to all manner of construction, the demands for cement and of the aggregate with which it is mixed have grown and grown. This has mainly manifested itself in the prodigious growth of demand for gravel and sand, now of the order of 100m. tons per annum, and it would be folly to pretend that the use of granite, limestone, and slag as a concrete aggregate forms more than a minor contribution to the construction industry's requirements.

The Group's stake in gravel and sand is comparatively small, except in the Scottish industrial belt, in Co. Durham, and in Cumberland. It is an industry in which profits are hard to come by except by economies of scale, and we have seen in the Tarmac narrative an abandonment of gravel projects after a short experience.

The main interest in my research has been to ascertain why the Group's quarried stone and slag has had as much success in the concrete market as has in fact been the case. First, however, we must note that the concrete market divides itself into two parts. We have the pre-cast concrete goods industry, slabs, kerbs, claddings, and the like. Secondly, concrete made and poured *in situ*. More and more the latter business has become that of 'ready-mix' concrete, mixed primarily at a central depot, and continuously mixed in transit to the job. The development of ready-mix concrete, which will always be associated with its Australian pioneers, has been very speedy over the last decade, and is still growing.

In both parts of the market, quality of aggregate has been of importance. Many consulting engineers and architects have specified granite aggregate in particular for certain jobs, and Cliffe Hill is an outstanding example of a quarry which has benefited from this. Middleton-in-Teesdale whinstone was exclusively specified by one of the great banks for concrete associated with their vaults and safes. While the iron and steel industry have used slag as a concrete aggregate in heavy and most exacting construction, the efforts of the slag industry to popularize slag as a concrete aggregate have met with little success.

To the student of economics, however, it is the other factor that is more interesting. Gravel is not everywhere available in Great Britain, and what is more, river gravel arises in thin beds, of the order of 15 feet of thickness. This is where limestone, the aggregate next in cheapness of production to gravel, figures in the developing story. For instance, there is no local sand and gravel in the Bristol area, apart from that dredged from the Bristol Channel. The concrete and mortar requirements of the conurbation have therefore had to be satisfied by limestone, and they have always formed a significant part of the sales of Crow Catchpole's quarry at Cheddar. The narrative of the Fewsters' business on Tyneside shows how the exhaustion of gravel in that river valley has led to the highly successful North Tyne limestone venture in conjunction with A.R.C. The gravel pits of West London are largely a thing of the past, as a result of exhaustion and town planning. Even in the Midlands, the Group's limestone quarries have now a respectably-sized trade in this market. It is perhaps more appropriate to my conclusion to this history to consider the further effect on this market of the recent sensational increases in the cost of road haulage of bulk materials.

3. Railtrack Ballast

This market has been supplied from the earliest days of this history. I have mentioned it as a step forward in the Cliffe Hill narrative in 1894, and there has been a continuous demand in granite, slag and limestone till the story closes. There have been considerable fluctuations for a variety of reasons. The policy of railway management has led to the cessation of quarrying on their own account, one important example of this being the leasing of Caldon Low to John Hadfield & Sons in 1934. The fact that only one size is required, generally $1\frac{1}{2}''$ to $\frac{3}{4}''$, creates problems of disposal of other sizes which the railways have found troublesome to solve. The general policy of closing down of unremunerative lines in the last fifteen years has reduced demand, while the introduction of diesel locomotives, and of continuously welded rails, having put heavier duty on to the ballast, thicknesses have been increasing from the old $6''$ up to as much as $1'\ 6''$ in some instances. Closing of lines, too, has upset supply arrangements, sometimes to the advantage and at other times to the disadvantage of the Group. Heavier duty has meant stiffer specifications, and this on the whole has helped our constituents.

While throughout Great Britain the Group is participating in this market, railway engineering policy has tended to canalize it in certain directions. Where, in past decades, one or other of the Group aggregates has been used on main lines, railway engineers insist on continuity, e.g. they will not permit mixing of granite and slag. Slag has never been acceptable to the London Underground Railways, with their third-rail low voltage systems, for fear of 'shorts' by the presence of particles of iron.

One way and another, the Group's participation has approximated to 10 per cent of the whole of this market, which runs into millions of tons per annum.

4. Pipes and Bricks

Though the Division's interest is limited to a pipe plant at Craig-Lelo and a small brickworks at Hillhead quarry, it is to be noted that both were started by Hillhead Hughes for the specific purpose of using quarry products.

5. Non-ferrous Mining

The successful activity in this field is that of the production and marketing of fluorspar by Derbyshire Stone. As stated in the narrative, this industry was entered in 1940, and the mine at Masson Hill, near Matlock, was taken over, and originally mined by square-set stoping, a method then new to this country but extensively employed in Canada. Consequently the Ministry of Supply obtained the services of six Canadian miners from the Royal Canadian Engineers to get things going.[1] Fortunately, the ore-bed outcrops and open-cast working was rapidly developed. Other deposits were acquired and a considerable business built up in metallurgical fluorspar. The increasing demand for both acid-grade and metallurgical fluorspar brought many people into the Derbyshire field, including Laportes, who were concentrating on the acid grade. When Derbyshire Stone installed a flotation plant in 1957 a clash between the two big producers became possible, but it was averted in 1963 by the good arrangement which still prevails, whereby Derbyshire Stone abandoned mining but retained the sales of the metallurgical product. The industry, with a considerable export factor, continues to grow, and an excellent and comprehensive account of its history and prospects has recently been published.[2]

The history of mining, by its very nature, includes more failures

[1] See *The Iron and Coal Trades Review*, 8 Oct. 1943.
[2] Dr. B. L. Hodge, The U.K. Fluorspar Industry and its Basis, *Industrial Minerals*, April 1970.

than successes, and I must record the failure of Crow Catchpole in barytes mining in Co. Durham after World War II, of Fewsters in the same field in Cumberland, and of Derbyshire Stone in conjunction with Johannesburg Consolidated Investments in deep mining for lead at Riber in Derbyshire in the fifties.

The story of Ord & Maddison's coal ventures last century is now lost, but a successful minor operation in open-cast coal mining has been carried on in Co. Durham as a result of the acquisition of Taylors (Crook) Ltd.

6. Foamed and Granulated Slag

Slag is a valuable mixture of minerals, and has the unique property of being primarily available in a molten state. Consequently, there are attractive possibilities of uses of slag other than letting it cool in the atmosphere to compete in the aggregate market with natural stone. The Group's slag constituents have been involved, normally in supply of slag or in research, in many of these alternatives, among which may be mentioned scoria bricks, slag wool, and slagceram. Some 40–50,000 tons per annum of slag wool are produced by the Group for the insulation trade. The only other lines of significant economic importance to this history are those which form the title of this sub-heading.

Foamed slag aggregate has been produced in Great Britain since 1935. Until comparatively recently its use, and that of other lightweight aggregates has been largely confined in the United Kingdom to the manufacture of building blocks. During the sixties, however, the results of much research, mainly by the Building Research Station, into reinforced lightweight concrete have been published and there are now some notable examples of far more sophisticated use of foamed slag than in mere blocks.

The Group's interest is confined to its one-third share in the plant at Consett, acquired in 1955 by Tarslag (q.v.).

When molten slag is cooled rapidly by high pressure water jets, crystals have no time to form and it solidifies as a glassy-type material known as granulated slag because it takes the form of small granules. It has marked hydraulic setting properties, and is

The Lion Carving

HOPTON WOOD—PAST AND PRESENT:

Middleton Limestone Mine *Overleaf*

used largely on the Continent in the manufacture of cement. At the acrimonious meeting at the Ministry of Transport in 1947, referred to in Chapter VI, Col. D. W. Cleaver of A.R.C., the spokesman for the quarrying interests, suggested that the slag industry should solve its problems by a massive entry into this market. In fact, this has never taken place, partly by reason of the strength of the British Portland Cement interests, and partly because the slag industry found in the expansion of coated macadam a far more profitable outlet. Nevertheless, for many years to the close of this history, Tarmac have exported granulated slag from Scunthorpe to Lübeck in the Baltic for this purpose, though, subsequently economics has brought this particular business to an end. Quantities reached 100,000 tons per annum.

Yet granulated slag may still be of interest to the Group. It is very popular in France, mixed with coarse aggregate, in road bases, because of its cementing properties.

7. Filter Media

The function of the medium in a percolating filter is to provide an extended surface to serve as a support for the organisms, including bacteria and fungi, which brings about the purification of polluted liquids by the filter.[1] Though granite has been extensively used—the Cliffe Hill narrative records an important contract as far back as 1926–7—the vesicular nature of the surface of slag is of great value in the proper operation of a biological percolating filter-bed. Mr. J. H. Haiste, the well-known consultant in this field, stated[2] in 1960 that slag had been used in Great Britain for fifty years with complete satisfaction in sewage disposal works and that approximately a million tons had been so used in the last five years. Tarmac, in conjunction with the specialist merchants in this field, Messrs. Summerfield & Lang of Liverpool, have paid considerable attention to this market, including the formation in 1962 of a small company to undertake spreading of the material.

[1] See British Standard Specification 1438 of 1948.
[2] Foreword to British Slag Federation pamphlet of 1960.

MRM

8. *Metal Reclamation*

We have seen in Chapter V to what a great extent the iron and steel industry has abandoned direct participation in the processing and marketing of blast-furnace slag. Similarly the industry has largely handed over to specialist entrepreneurs the recovery of steel from the slag produced from steel cupolas. Not only are the engineering problems of extraction of considerable complexity, but the organization involved does not fit into the pattern of three-shift continuous working, and this produces labour difficulties for steelworks managers. They have also been baffled by the special difficulties of disposal of the steel slag after the metal has been extracted from it. Often it must weather for as long as eighteen months. Its high specific gravity compared with blast-furnace slag has handicapped its marketing. There has therefore been in existence for a very long time a metal reclamation industry ancillary to steelmaking, and it is unfortunate for the Tarmac Derby Group that Tarmac's entry into this activity dates back only to 1958 with a plant at Skinningrove. Even so, metal reclamation has proved a thoroughly rewarding affair. Tarslag started it tentatively at Turners Hill in the early fifties, and a proper plant was installed at Round Oak in 1959. At Workington, though the exploitation of the ferro-manganese slag has not succeeded, the metal reclamation plant earns a proper return on capital, while the operations started in 1963 at Park Gate, and described at some length in the Prestwich and Tarmac narratives (q.v.) have resulted in a very useful business. It will be noted that the Park Gate agreement included the grinding of phosphoric slag for fertilizer. Though this particular activity has had to be abandoned owing to the low P_2O_5 content of the Park Gate basic slag, it is an example of the minor activities which steel-works management tends to entrust to the metal extraction contractor with his specialist plant and labour force.

The fact that the Division has only a comparatively small participation in this industry would seem to flow from an historical approach upon which I shall be commenting in my conclusion to this book.

Conclusion

Experience teaches slowly, and at the cost of mistakes

J. A. Froude

The history ends in mid-1968, but it will not be published till 1971. I could conclude it in the light of the excellent results, in every department, of 1967. This would, I think, be unrealistic. The last three years have been disappointing. I suggest that, once again, the age-long feud between road-user and road-maker, to which I referred in Chapter I, has raised its head. Mrs. Barbara Castle, with her controversial Transport Act of 1968, and her successors at the Ministry of Transport with their short-sighted cuts in road maintenance seem to me in the long line of succession to King James I in 1621, to the Local Government Board at the turn of this century—so castigated by those saints of the Socialist hagiology, the Webbs, and to Sir Winston Churchill in 1926. They all seem to have shared the fond belief that, if one makes it difficult, expensive, and unpleasant to use the roads, then the growth of traffic will be restrained. It never has worked, and I do not believe that it will in the last thirty years of the twentieth century. Modern administrations have not their predecessors' excuse that the future is all unknown. It is folly to ignore the predictions of economics and sociology, and they combine to tell us that the automobile age is still expanding and that motor vehicle population will go on increasing rapidly for decades.

Nobody except visionaries contemplate roads without aggregate and so I believe that roadstone continues to be a growth industry. Looking at the narrative of Tarmac Ltd. alone, there have been already two 'plateau' periods, 1927–34 and 1954–8, in the last half-century of great development, and I think that the present time is simply another one.

Though I have recalled the shadowy beginnings of this great business in the eighteenth century, and its more positive roots in the latter part of the nineteenth, we can, I think, draw no practical lessons from the past till we come to Purnell Hooley's 'invention' of Tarmac. It is proper to call Hooley an inventor, though, like nearly all the great industrial inventors, he had fore-runners. Like most of them, the financial benefit of his invention was not reaped by himself, but by his patron and the latter's successors. Men like Hooley are rare, and a characteristic of first-class management is the ability to distinguish between them and the mass of mere talent which so often obscures them. Sir Alfred Hickman seems to have done just this.

The outstanding commercial success recorded in this book is the development of coated slag. Before I comment on it, my readers must forgive one personal note, and that is to say that, if they detect adverse criticism in some of my remarks in this conclusion it is largely due to my reflections on my own mistakes. The hero of this success story is Cecil Martin. To one who knew him well in his heyday, his greatest qualities were his industry, his integrity, and his single-mindedness of purpose. 'Follow the golden gleam' was a favourite expression of his, and his ability to do so flowed from the personal industry that created and maintained the necessary organization of production, transport, and sales manage-ment, and that integrity which was the secret of his successful negotiations with ironmasters, customers, and competitors. Tarmac's success, under the leadership of such a man, can cause no surprise. They were operating in a growth market, were, as a result of iron history well placed geographically, and the small number of sources of slag introduced that flavour of monopoly which is not the least desirable item in the recipe for commercial success.

Yet single-mindedness has its dangers. The gleam may dazzle the eye, and may inhibit the detection of paths other than that which it so brightly illuminates. Disappointment with the early results of diverse but allied projects certainly affected the judgement of the Tarmac Board in the case of gravel, road asphalt, and Scotland to the detriment of the business. Less excusable were the inertia and over-caution which stultified Crow Catchpole's growth. All the slag constituents seem to have missed the possibilities of metal reclamation till late in the story. There is no doubt that these errors of omission had to be paid for dearly in some instances, for example in the acquisition of Cliffe Hill.

As, for over twelve months, I have been gathering by reading, talking, and travelling, the material for Chapters III and VI, it has been more and more evident to me that the secret of successful quarrying is size, that size involves great expenditures of capital, and that in this industrial activity one can agree with Voltaire that *Dieu est toujours pour les gros bataillons.* Consider the case of Rowley Regis. In seventy years nothing has fundamentally changed physically; there is the great deposit of first-class granite situated on the edge of a vast conurbation. Yet the £1 shares issued at par in 1900 only fetched 26s. in 1951, despite the decline in the value of money. The resources to develop were not available. Then think what Tarslag and Tarmac have made of this remarkable property by their command of both capital and skills.

The leading concern in my story in this field is of course Derbyshire Stone, and it is undoubtedly John Hadfield who shares with Cecil Martin the credit for the fundamental construction which enables the Roadstone Division of Tarmac Derby to occupy the position in the country's economy which it fills today. One cannot but admire his fertility of mind, the imagination which grasped the importance of the principle of size in quarrying, the modernity of his ideas in labour relations, price structure, and marketing. Yet in 1959, Derbyshire Stone and Tarmac, both able and anxious to expand on a big scale, and having shared the frustrating experiences of the roadstone industry in the late fifties, took very different measures to achieve their object. Derbyshire Stone pursued a policy of complete diversification. Tarmac, on the contrary, greatly increased their stake in roadstone.

Only ten years afterwards it would be presumptuous and indeed stupid to pass any final judgement on these two contrasting policies. My diffident, and very provisional, comment amounts to this. Tarmac were right in their confidence in the growth of the roadstone industry, but probably over-estimated its rate of expansion maybe, therefore, paying an excessive price for some of their acquisitions. Derbyshire Stone were right in broadening the base of their business, but failed to appreciate fully the dangers of lack of intimate understanding of industries of which their management had no previous experience. It is noteworthy that if in 1959 Tarmac were wrong in over-estimating and Derbyshire Stone in under-estimating roadstone growth, they both within a very few years took steps calculated to correct their mistakes. Tarmac's imaginative vertical expansion into the bitumen business is not dependent for success on any particular rate of roadstone growth. Derbyshire Stone's horizontal expansion into Kings in 1965 showed a quick reaction to increasing demand in an area remote from their previous operations.

To one unfamiliar with the 'black-top' section of the road industry, the greatest surprise to be derived from reading my account must surely be the inferiority of the commercial success of road asphalt to that of coated macadam. It seems all wrong. Research and technological control have been poured into asphalt for eighty years. Hit-and-miss methods have been very frequently the best description that can be applied to its rival's techniques. Hot asphalt is the darling of leading road engineers; every black-topped motorway in Britain presents to the motorist a surface of hot asphalt almost always most excellent in appearance and in riding qualities. Moreover, statistics indicate a greater rate of growth in asphalt than in coated macadam. Why then, for example, in great years like 1964 and 1967, would it appear that the profits derived by the Division's constituents from cold asphalt, a minor if important part of its coated macadam trade, can look in the face the whole of the hot asphalt profits? Of course, asphalt is dearer than coated macadam. Of course, by its nature, it cannot participate in that vast field of small miscellaneous sales that is the backbone of the latter's trade, but there must be some more fundamental reason than these.

My belief is that the answer lies in the superior price structure of the coated macadam industry, at any rate over the years since World War II. I must offer an explanation of how this superiority has come about.

Coated macadam is based on static plants, themselves based on sources of aggregate. These sources of aggregate are limited, especially in an era of planning and of steel industry concentration, and the possession of them confers on their owners a stability which is quite distinct from the technical and administrative skills which they may share with their asphalt counterparts. We have seen what large amounts of capital have been expended by the leading constituents of the Group in order to secure this stability for the future. The commercial manager of a modern coated macadam concern therefore knows who are his effective competitors. He may be irritated and distracted from time to time by the gadfly attacks of newcomers, but his sense of proportion should preserve him from panic.

His opposite numbers in the trade share his knowledge. They know, too, that the size of the market is controlled by the Government's allocation of the national resources, and that cutting prices will not increase it to any significant extent. They share with him appreciation of the fundamental geological and geographical facts which I set out in the Introduction. The stage is set for a profitable sharing out of the available trade, and Chapter IX shows, how, since the thirties, the psychological atmosphere has been such that this is the way that the drama has proceeded, whether by conscious agreement or by price leadership.

Asphalt is commercially quite a different type of business. The raw materials are the appropriate aggregates nearest to the particular job, and bitumen, and these are normally available to all tenderers. Originally, as this history shows, the possession of natural rock asphalte was vital. The only survivor of this particular advantageous position is Limmer & Trinidad, with their interest in Trinidad lake asphalte, and our story has shown that this has diminished substantially in importance. The Tarmac Derby Group, too, has gained an advantage by being able, by virtue of the size of its consumption of bitumen, to enter the residual bitumen industry itself. Neither of these cases alters the fact that the road asphalt business is really a specialized branch of civil engineering. From this can be inferred that the unexpected competitor is always a danger, making a common price policy a difficult matter, and that success flows fundamentally not from possession of raw material, but from skill in choice of job, in the detail of tendering, above all in a correct selection of top, middle, and local management, and of plant.

In any case, it is far easier to establish the prices of specified commodities than the prices of contractors' jobs, which, almost always, exhibit some features to which one tenderer will devote more attention than will another.

The development of sales of aggregate in the non-roadmaking market is a feature of the last years of my survey of the Division's history. The growth in demand is as certain as any economic projection can be; it has been recently estimated at 5 to 6 per cent per annum for the remainder of the century. Already, in this history, we have observed the rapid exhaustion of some gravel supplies and the substitution of limestone, of which the Division has large reserves in the north of England, in Derbyshire, and in North Wales. The new feature seems to be the sensational increase in the cost of bulk transport, following the political and social policies of the Transport Act 1968. I have already said that I do not believe that this measure will affect the long-term demand for roadstone, and, I would add, for aggregate generally, but it must alter the pattern of distribution. In addition to some exhaustion of supply, the gravel industry may now face some geographical restriction.

As to the other non-roadmaking activities, in particular industrial limestone and metal reclamation, their profitability can only suggest increased effort.

In this century of growth, great success has been often marred by mistakes. I have recorded errors of omission and commission. Inertia, excess of caution, loss of heart, denial of capital to temporarily ailing branches, missed opportunities are all discernible together with excessive optimism and light-hearted plunging into the unknown. My final remarks must emphasize the importance of adequate planning. Management skills in planning are available of a nature and degree largely unknown, and certainly in some instances unused, ten years ago. I can laugh, for instance, at my own old ideas of cost accountancy. Much of the market research of the past was superficial, and in particular paid far too little attention to the all-important factor of price. Acquisitions priced merely on past accountancy, investigations of title, and proof of physical reserves carry the potentiality of danger, particularly when they are in regions of which the buyer has little knowledge. To suggest that all risks can be eliminated, and growing prosperity assured by the application of the planning skills now available would be, of course, arrant nonsense. It has been said that it takes ten years to make the idea of a five-year plan work satisfactorily. What I am suggesting is that one great lesson to be learnt from these pages is that the hard work and imagination, sometimes in the form of 'hunch', which have brought about the success that they narrate, should be applied to seeing that future planning is no routine affair, but is vivid in every aspect.

A Brief Chronology

18th century Neuchatel asphalte deposits discovered. Hopton Wood, Caldon Low, and Bankfield (Clitheroe) in operation.

ca. 1850 Ord & Maddison started.

1865 Crows, Catchpoles and Briggs distilling tar.

1869 John Hadfield & Sons started.

1873 The Neuchatel Asphalte Co. Ltd. registered.

ca. 1880 Prestwich, and Chittenden and Simmons started.

1891 Fitzmaurice acquired Cliffe Hill.

1895 Constables acquired Cawdor, Matlock.

1898 (New) Northern Quarries at Silverdale; Cuthbert acquired Kings.

1900 Rowley Regis Granite Quarries Ltd. registered.

1903 Tarmac Ltd. registered.

1909 H. V. Smith registered.

1913 Highways Construction started; Majors at Stockton.

1915 Ragusa started.

1920 Tarslag incorporated.

1921   Hughes Bros. started.

1923   Crow Catchpole started in coated macadam.

1925   Silvertown incorporated.

1934   Taylors (Crook) Ltd. started.

1935   Derbyshire Stone merger; Hillhead Hughes incorporated.

1945   Fewsters started quarrying.

1951   Tarslag acquired Rowley Regis.

1952   Ragusa acquired Chittenden & Simmons.

1957   Amasco merger.

1959   Tarmac acquired Crow Catchpole and Tarslag.

1960   Tarmac acquired Taylors (Crook).

1962   Amasco acquired H. V. Smith. Tarmac acquired Ord & Maddison.

1963   Tarmac acquired Prestwich. Derbyshire Stone and Neuchatel acquired Amasco.

1964   Tarmac acquired Fewsters' quarrying interests and New Northern Quarries.

1965   Tarmac acquired Cliffe Hill and Hillhead Hughes. Derbyshire Stone acquired Kings. Derbyshire Stone acquired Neuchatel, and thereby the whole of Amasco.

1968   Tarmac Derby merger. Acquisition of Silvertown.

Bibliography

CHAPTER I

1. *Lives of the Engineers.* Smiles, John Murray, 1861.
2. *English Local Government.* S. and B. Webb, Longmans, Green, 1913.
3. *The Story of the Road.* J. W. Gregory, Alexander Maclehose, 1931.
4. *The King's Highway.* Rees Jeffreys, Batchworth, 1949.
5. *Britain between the Wars.* C. L. Mowat, Methuen, 1955.
6. *Highways.* C. A. O'Flaherty, Arnold, 1967.
7. 'Roads, a New Approach'. British Road Federation.
8. 'Basic Road Statistics'. British Road Federation, 1969.

CHAPTER V

1. *Capital Development in Steel.* Andrews and Brunner, Blackwell, 1951.
2. *Roman Roads in Britain.* Vol. I, I. D. Margary, Phoenix House, 1955.
3. *History of the British Steel Industry.* Carr and Taplin, Blackwell, 1962.
4. *Victorian Cities.* Asa Briggs, Odhams, 1965.
5. 'The Black Country Iron Industry'. Gale, Iron & Steel Institute, 1966.

6. *The Economic History of the British Iron and Steel Industry,* *1784–1879.* Birch, Cass, 1967.
7. 'The Development of Corby Works'. Sir F. Scopes, Stewarts & Lloyds Ltd., 1968.

CHAPTER VI

1. *Practical Stone Quarrying.* Greenwood and Elsden, Corsby Lockwood, 1913.
2. 'Sources of Road Aggregate in Great Britain'. Road Research Laboratory, H.M.S.O., 1960.
3. (Unpublished) *One Hundred Years of Stonebreaking,* 'Macadamer', 1962.

CHAPTER VII

1. 'Bituminous Materials in Road Construction'. Road Research Laboratory, H.M.S.O., 1961.
2. *Asphaltic Road Materials.* Hatherley and Leaver, Arnold, 1967.

CHAPTER VIII

1. *Motor Bodywork.* H. J. Butler, W. R. Howell & Co., 1924.
2. The F.C.M.I. Lectures, 1957 (J. B. F. Earle).
3. *An Economic History of Transport.* C. I. Savage, Hutchinson, 1959.
4. 'Hydraulics at Work'. Golden Jubilee Issue, Edbro Ltd., 1969.
5. 'English Foreign Trade in Roadstone'. J. B. F. Earle, *National Westminster Bank Quarterly Review,* May 1970.

CHAPTER IX

1. *The Trust Movement in British Industry.* H. W. Macrosty, Longmans, Green, 1907.
2. 'The Restrictive Trade Practices Acts, 1956 and 1968'. H.M.S.O.

CHAPTER X

1. The F.C.M.I. Lectures, 1957 (James Drake).
2. *Industrial & Agricultural Uses of Limestone.* J. J. Berliner & Co., New York City.
3. 'Natural Chalk Whiting'. Welwyn Hall Research Association, 1969.
4. 'The U.K. Fluorspar Industry and its Basis'. Dr. B. L. Hodge, *Industrial Minerals*, April 1970.

Indexes

INDEX OF WORKS, QUARRIES AND SLAGTIPS MENTIONED
* Active, mid-1968

GENERAL INDEX

INDEX OF CONSTITUENT AND ASSOCIATED COMPANIES
† Detailed narratives

369 0123481

612.67

DATE DUE

Demco, Inc. 38-293